Norma

D1327350

I hope you will find this both interesting and useful.

Dave

Data Monitoring Committees in Clinical Trials

A Practical Perspective

Statistics in Practice

Advisory Editor

Stephen Senn
University College London, UK

Founding Editor

Vic Barnett
Nottingham Trent University, UK

Statistics in Practice is an important international series of texts which provide detailed coverage of statistical concepts, methods and worked case studies in specific fields of investigation and study.

With sound motivation and many worked practical examples, the books show in down-to-earth terms how to select and use an appropriate range of statistical techniques in a particular practical field within each title's special topic area.

The books provide statistical support for professionals and research workers across a range of employment fields and research environments. Subject areas covered include medicine and pharmaceutics; industry, finance and commerce; public services; the earth and environmental sciences, and so on.

The books also provide support to students studying statistical courses applied to the above area. The demand for graduates to be equipped for the work environment has led to such courses becoming increasingly prevalent at universities and colleges.

It is our aim to present judiciously chosen and well-written workbooks to meet everyday practical needs. Feedback of views from readers will be most valuable to monitor the success of this aim.

A complete list of titles in this series appears at the end of the volume.

Data Monitoring Committees in Clinical Trials

A Practical Perspective

Susan S Ellenberg

Thomas R Fleming
University of Washington, USA

David L DeMets
University of Wisconsin-Madison, USA

JOHN WILEY & SONS, LTD

Copyright © 2003 John Wiley & Sons Ltd, The Atrium, Southern Gate, Chichester, West Sussex PO19 8SQ, England

Telephone (+44) 1243 779777

Email (for orders and customer service enquiries): cs-books@wiley.co.uk
Visit our Home Page on www.wileyeurope.com or www.wiley.com

Reprinted April 2003

All Rights Reserved. No part of this publication may be reproduced, stored in a retrieval system or transmitted in any form or by any means, electronic, mechanical, photocopying, recording, scanning or otherwise, except under the terms of the Copyright, Designs and Patents Act 1988 or under the terms of a licence issued by the Copyright Licensing Agency Ltd, 90 Tottenham Court Road, London W1T 4LP, UK, without the permission in writing of the Publisher. Requests to the Publisher should be addressed to the Permissions Department, John Wiley & Sons Ltd, The Atrium, Southern Gate, Chichester, West Sussex PO19 8SQ, England, or emailed to permreq@wiley.co.uk, or faxed to (+44) 1243 770620.

This publication is designed to provide accurate and authoritative information in regard to the subject matter covered. It is sold on the understanding that the Publisher is not engaged in rendering professional services. If professional advice or other expert assistance is required, the services of a competent professional should be sought.

Other Wiley Editorial Offices

John Wiley & Sons Inc., 111 River Street, Hoboken, NJ 07030, USA

Jossey-Bass, 989 Market Street, San Francisco, CA 94103-1741, USA

Wiley-VCH Verlag GmbH, Boschstr. 12, D-69469 Weinheim, Germany

John Wiley & Sons Australia Ltd, 33 Park Road, Milton, Queensland 4064, Australia

John Wiley & Sons (Asia) Pte Ltd, 2 Clementi Loop #02-01, Jin Xing Distripark, Singapore 129809

John Wiley & Sons Canada Ltd, 22 Worcester Road, Etobicoke, Ontario, Canada M9W 1L1

Wiley also publishes its books in a variety of electronic formats. Some content that appears in print may not be available in electronic books.

British Library Cataloguing in Publication Data

A catalogue record for this book is available from the British Library

ISBN 0-471-48986-7

Typeset in 10/12pt Photina by Laserwords Private Limited, Chennai, India
Printed and bound in Great Britain by TJ International, Padstow, Cornwall
This book is printed on acid-free paper responsibly manufactured from sustainable forestry in which at least two trees are planted for each one used for paper production.

Contents

Preface

The randomized clinical trial has been recognized as the gold standard for evaluation of medical interventions for only half a century (Doll, 1998). Over the past several decades, the increasingly central position of randomized clinical trials in medical research has led to continual advances in the development of methodology for the design, conduct and analysis of these studies. An enormous body of literature relating to clinical trials methodology is now available, a professional society focusing on clinical trials has been established (Roth, 1980; www.sctweb.org), and a large number of statisticians, clinicians and epidemiologists consider clinical trials as their primary area of research and/or application.

One area of clinical trials that has received relatively little attention but that can be critical to the ethics, efficiency, integrity and credibility of clinical trials and the conclusions of such trials is the process of interim monitoring of the accumulating data. To an increasing extent, interim monitoring is becoming the province of formally established committees. While a great deal has been written about statistical methods for interim data monitoring, the practical aspects of who should serve on data monitoring committees (DMCs) or otherwise be involved in the monitoring process, what data should be monitored and how frequently, and what are the necessary and appropriate lines of communication have received limited discussion. Since DMCs are given major responsibilities for ensuring the continuing safety of trial participants, relevance of the trial question, appropriateness of the treatment protocol, and integrity of the accumulating data, it is important to understand the ways in which these committees meet such responsibilities.

A word about terminology. Committees to monitor accumulating data from clinical trials go by a variety of names. The two most frequent of these are probably 'data and safety monitoring board' and 'data monitoring committee', but there are many other variations (Ellenberg, 2001). We have arbitrarily selected 'data monitoring committee', in part because of its simplicity and in part because this is the term used by international regulatory authorities (www.ifpma.org/ich1.html).

From time to time, papers describing the experience of particular DMCs, as well as papers addressing general approaches for operating and serving on such committees, have been published; a number of these are referenced in Chapter 1. These papers have provided some valuable insights into the monitoring process. In

1992 an international workshop was held at the National Institutes of Health to discuss different approaches to data monitoring that had been or were being used in a variety of settings, and the proceedings were published as a special issue of the journal *Statistics in Medicine* (Ellenberg *et al.*, 1993). At this workshop, individuals with substantial practical experience in interim data monitoring reported on their preferred operating models, and there was substantial discussion of the advantages and disadvantages of the different approaches presented. Up to now, those workshop proceedings plus the aforementioned papers have constituted the primary references for those interested in learning about the various operating models in use for DMCs, as well as the diversity of issues these committees may consider.

The use of DMCs has continued to grow, especially with respect to trials sponsored by pharmaceutical companies. The demand for individuals to serve on these committees is high; it is increasingly difficult to ensure that any DMC will include at least some members with prior experience on other DMCs. As individuals with extensive experience coordinating and/or serving on such committees, the authors of this book are frequently asked for advice concerning their operation (from trial organizers/sponsors) and the scope of responsibilities of committee members (from new members of such committees). The increasing interest in these issues led us to believe that a comprehensive reference on the practice of interim data monitoring and the structure and operation of DMCs was needed; that was our primary motivation for writing this book.

The book is intended for those involved with or otherwise interested in the clinical trials process. We expect this group will include statisticians, physicians and nurses, trial administrators and coordinators, regulatory affairs professionals, bioethicists, and patient advocates. The issues are relevant to trials sponsored by government funding agencies as well as by pharmaceutical and medical device companies, although approaches taken may differ in different contexts. We also believe this book should be of interest to those involved in the evaluation and reporting of trial results – for example, medical journal editors and science journalists for lay publications – as the process of trial monitoring has important implications for the interpretation of results. We have attempted to keep the material non-technical, so as to make it accessible to as large a part of the clinical trials community as possible.

Every chapter in the book addresses an issue that has been debated among those with DMC experience in different settings. Our intent is to describe the issues clearly as well as to describe the arguments that have been made for and against different approaches that might be taken. We will identify areas where there appears to be a general consensus, and occasionally recommend a particular approach even when there is no widespread consensus on that issue. For the most part, however, our goal is to clarify the types of decisions that must be made in implementing DMCs and not to provide a prescription for their operation. There is no 'one size fits all' for DMCs; different models may be needed for different situations.

We begin with some introductory background and some historical notes on the use of DMCs in different contexts. Next, we address the scope of responsibilities that may be assigned to a DMC. Some committees are charged with reviewing outcome data only (or even safety data only); others are asked to review the initial protocol, monitor the conduct of the study by assessing accrual, eligibility, compliance with protocol, losses to follow-up, and other issues that are ultimately relevant to the value and credibility of a trial. The specific responsibilities delegated to a committee monitoring a particular trial will influence other operational aspects, such as committee composition.

In Chapter 3 we consider the committee membership: what types of expertise should be represented on all committees, other relevant factors in selecting committee members, optimal committee size, methods of selecting committees (and committee chairs). An important issue regarding committee membership that we discuss in some detail is conflict of interest.

Chapter 4 continues the consideration of conflicts of interest in the broader context of the independence of the committee. We discuss what is meant by an 'independent' committee, and the potential consequences for the trial and its credibility when the committee's independence is called into question. We also discuss the various types of trials for which independence of the DMC may be most critical.

Chapter 5 deals with one of the most controversial issues relating to the interim monitoring of clinical trial data: the extent to which any interim data, and unblinded interim data in particular, should be released to individuals or groups other than the committee itself. It has been argued that there may be a 'need to know' for some groups such as the sponsor or the regulatory authority; it has also been argued there is a 'right to know' for participating investigators, study subjects, and the general public. Others believe that limiting access to interim results is essential to the successful completion of clinical trials. This chapter focuses on such debates, and their potential implications for trial integrity.

In Chapter 6 we deal with the logistical issues – how often a committee should meet, how long the meetings need to be, how they are conducted, the content of the report the committee is to consider, the preparation and content of meeting minutes, and a number of other issues. Many groups who regularly sponsor and/or coordinate clinical trials have developed their own approaches to these issues, but these approaches can be quite different, even for similar types of clinical trials. Some might consider these types of issues part of the 'minutae' of clinical trials; our experience, however, is that the quality and reliability of the monitoring process may depend very heavily on just these types of issues.

Chapter 7 addresses the very important but little discussed topic of how the DMC interacts with other trial components. There are many constituencies involved in any given trial, including the sponsor(s), the investigators, the statistical coordinating center, the study steering committee, the institutional review board(s), and of course the patients. There is also a variety of modes of interaction, both formal (e.g., submitting reports) and informal (e.g., attending meetings of other components where unstructured discussion may take place).

Chapter 8 provides an overview of the various statistical approaches for interim monitoring of clinical trial data, and some discussion of why some approaches may be more useful in some circumstances than others. In this chapter, we also discuss the rationale for using these statistical tools in the monitoring process, as they have been widely but not universally adopted by DMCs. This discussion includes consideration of the different philosophies that have been expressed regarding the appropriateness of stopping clinical trials before they have collected all the information that was specified at the outset, a discussion that of necessity brings in the ethical issues that have been brought to bear on this determination.

In Chapter 9 we consider in more detail the monitoring approaches best suited to different types of trial, and describe an alternative to an independent monitoring committee that has been found useful in some settings.

Finally, in Chapter 10 we review regulatory considerations that may affect the operation of a DMC. There is very little in the US Code of Federal Regulations concerning DMCs; they are certainly not mandated except in one very limited circumstance. But there are aspects of the regulatory process that are important for DMCs to be familiar with, and there have been occasions when interactions between regulatory authorities and DMCs have occurred. Such interactions raise important questions about where certain responsibilities may optimally reside. Shortly before this book went to press, the Food and Drug Administration issued a draft guidance document on the establishment and operation of DMCs, and that document is briefly summarized.

The reader will find real-life examples throughout the book. Many of these examples come from the direct experience of the authors and have not been written about previously; others have been described in prior publications. We hope these examples will demonstrate the types of decisions and dilemmas DMCs frequently face, and the consequent difficulty of establishing a set of fixed rules for the operation of these committees. Our goal with this book is to assist those who establish DMCs, those who serve on them, those who are participating in trials and depending on their judgment, as well as those who read, interpret and use the results of clinical trials.

The book has benefited enormously from the constructive advice of those who graciously agreed to read drafts and provide comments. Baruch Brody, Lawrence Friedman, Alan Hopkins, Desmond Julian, James Neaton, Stuart Pocock, David Stump and Janet Wittes reviewed drafts of most chapters and their input led us to make many improvements. Robert Temple, Jay Siegel, Scott Emerson, Tom Louis, Paul Canner and Jonas Ellenberg provided extremely helpful input on specific chapters. Diane Ames assisted in producing many of the figures. Sue Parman coordinated much circulation of material, arranged meetings and teleconferences, and assisted with the preparation of several chapters.

Thanks are also due to Helen Ramsey of Wiley, who encouraged the development of this book, and to Wiley editors Sharon Clutton, Siân Jones and Rob Calver for their assistance and collegiality throughout the process. We also appreciate the work of Richard Leigh, our copy editor, for the many modifications he suggested and queries he raised that improved the flow of the book and eliminated errors and ambiguity.

We are indebted to all our colleagues with whom we have served on DMCs, with whom we have worked in preparing reports to DMCs, and who have served on DMCs to which we have reported. Whatever value there may be in these pages derives from the fundamentally collaborative experience of monitoring clinical trial data and the mutual learning that ensues.

We would like to acknowledge partial support from National Institutes of Health grants NIHR37AI129168 (T.F.) and NIHR01CA18332 (D.D).

Finally, we are particularly grateful for the forbearance and support of our families – particularly our spouses, Jonas, Joli and Kathy – during the process of writing, rewriting, arguing, negotiating, and nitpicking as we made our way to the final manuscript.

REFERENCES

Doll R (1998) Controlled trials: the 1948 watershed. *British Medical Journal* **317**: 1217–1220.

Ellenberg SS (2001) Independent monitoring committees: rationale, operations and controversies. *Statistics in Medicine* **20**: 2573–2583.

Ellenberg SS, Geller N, Simon R, Yusuf S (eds) (1993) Proceedings of 'Practical issues in data monitoring of clinical trials', Bethesda, Maryland, USA, 27–28 January 1992. *Statistics in Medicine* **12**: 415–616.

Roth HP (1980) On the Society for Clinical Trials. *Controlled Clinical Trials* **1**: 81–82.

1

Introduction

Key Points

- The purpose of data monitoring committees (DMCs) is to protect the safety of trial participants, the credibility of the study and the validity of study results.
- DMCs have a long history in trials sponsored by government agencies in the USA and Europe.
- Pharmaceutical companies are increasing their use of DMCs in trials of investigational drugs, biologics and medical devices.
- Statistical methods have been developed for interim monitoring of clinical trials.
- While not all trials need DMCs, trials that address major health outcomes and are designed to definitively address efficacy and safety issues should incorporate DMC oversight.

1.1 MOTIVATION

In randomized clinical trials designed to assess the efficacy and safety of medical interventions, evolving data are typically reviewed on a periodic basis during the conduct of the study. These interim reviews are especially important in trials conducted in the setting of diseases that are life-threatening or result in irreversible major morbidity. Such reviews have many purposes. They may identify unacceptably slow rates of accrual or high rates of ineligibility determined after randomization, protocol violations that suggest that clarification of or changes to the study protocol are needed, or unexpectedly high dropout rates that threaten the trial's ability to produce credible results. The most important purpose, however, is to ensure that the trial remains appropriate and safe for the individuals who have been or are still to be enrolled. Unacceptable levels of treatment toxicity may require adjustment of dosage or schedule of administration, or even abandonment of the study. Efficacy results, too, must be monitored to enable benefit-to-risk assessments to be made. Interim results may demonstrate that one intervention group has such unfavorable outcomes with regard to survival or a major morbidity endpoint that its benefit-to-risk profile is clearly inferior to that of the comparator treatment. In such cases, it may be appropriate to terminate the inferior intervention or the entire trial early so that current study participants, as well as future patients, will no longer be provided the inferior treatment.

Relatively early in the development of modern clinical trial methodology, some investigators recognized that, despite the compelling ethical need to monitor the accumulating results, repeated review of interim data raised some problems. Repeated statistical testing was seen to increase the chance of a 'false positive' result unless nominal significance levels were somehow adjusted. In addition, it was recognized that awareness of the pattern of accumulating data on the part of investigators, sponsors or trial participants could affect the course of the trial and the validity of the results. For example, if investigators were aware that the interim trial results were favoring one of the treatment groups, they might be reluctant to continue to encourage adherence to all regimens in the trial, or to continue to enter patients on the trial, or they might limit the types of patients they would consider entering. Furthermore, influenced by financial or scientific conflicts of interest, investigators or the sponsor might take actions that could diminish the integrity or credibility of the trial. For example, a sponsor observing interim data showing that the new treatment had little if any effect on the prespecified primary endpoint but a much stronger effect on an important secondary endpoint might be tempted to switch the designation of these two endpoints.

A natural – and practical – approach to dealing with these problems is to assign sole responsibility for interim monitoring of data on safety and efficacy to a committee whose members have no involvement in the trial, no vested interest in the trial results, and sufficient understanding of trial design, conduct and data-analytical issues to interpret interim analyses with appropriate caution. These 'data monitoring committees' (DMCs) have become critical components of many clinical trials. The interim monitoring experience of an early AIDS clinical trial illustrates some of the inherent difficulties and challenges that are faced in reviewing the accumulating data from clinical trials.

Example 1.1: Treatment for HIV infection

Trial 002 of the Community Programs for Clinical Research in AIDS (CPCRA) was designed to compare the efficacy of two antiretroviral agents, zalcitabine (ddC) and didanosine (ddI), in HIV-infected patients who did not derive benefit from zidovudine (AZT), at that time the first-line treatment for HIV infection (Abrams *et al.*, 1994). When the trial was initiated, ddI was considered the first-line treatment in this patient population; the goal of the trial was to determine whether ddC was approximately equivalent to ddI by seeing whether as much as a 25% advantage for ddI in time to disease progression or death could be ruled out. A total of 467 patients were randomized to receive either ddI or ddC. To achieve the desired level of statistical power, it was calculated that patient follow-up would be needed until 243 patients had been observed to reach the endpoint of disease progression or death.

This trial was initiated in December 1990, at a time when little in the way of effective treatments for this population was available, when the numbers of new HIV infections and deaths were increasing, and when both the patient community and their physicians were increasingly desperate to identify treatments that could buy a little more time for those suffering from this disease. Patients entering such trials were generally young men who were facing a very premature death from a disease they may not have even known about at the time they contracted it. Further, more pharmaceutical companies were initiating drug development for treatment of HIV, but with a great deal of caution, as would be expected in a completely new disease area. While there are inherent tensions in all trials testing new agents for serious diseases, the atmosphere surrounding early trials of AIDS treatments, such as this one, was particularly 'high pressure'. Trial 002 was monitored by the DMC that had been established by the National Institute of Allergy and Infectious Diseases (NIAID) to oversee all of its extramural trials of treatment for HIV infection (DeMets *et al.*, 1995). The CPCRA was a clinical trials group funded by NIAID; therefore, access to interim data was limited to DMC members – none of whom were treating patients on this or any other NIAID-funded AIDS trial, or had any financial stake in the trial outcome – and to a limited number of NIAID staff.

The interim results from this trial, shown in Figures 1.1 and 1.2, illustrate how substantially relative risk estimates can change over time. At the first interim analysis in August 1991, the early trial results strongly favored ddI. At that time, the ddI group had experienced many fewer disease progressions (19 vs. 39) and fewer deaths (6 vs. 12) than the ddC group. The effects on laboratory markers were also more favorable in the ddI group. While the nominal p-value for the treatment difference in progressions at this analysis was an impressive 0.009, this value did not approach the protocol-specified early termination criterion at this early stage in the trial. The DMC considered these data as well as available information on toxicities and other relevant outcomes and recommended that the trial continue as designed.

As the figures show, the differences favoring ddI steadily disappeared over successive meetings of the DMC. At the final review, in August 1992, the DMC recommended that the study end as originally planned since the required number of events had been observed. The results at the end of the trial had shifted from strongly favoring ddI to showing a small advantage for ddC in this population. These data did provide strong statistical evidence that ddC was not inferior to ddI in the sense noted earlier.

Had the results from the initial interim analysis of the CPCRA 002 trial been broadly disseminated, it is most unlikely that the trial would have continued, given the urgent desire to identify optimal therapeutic approaches and

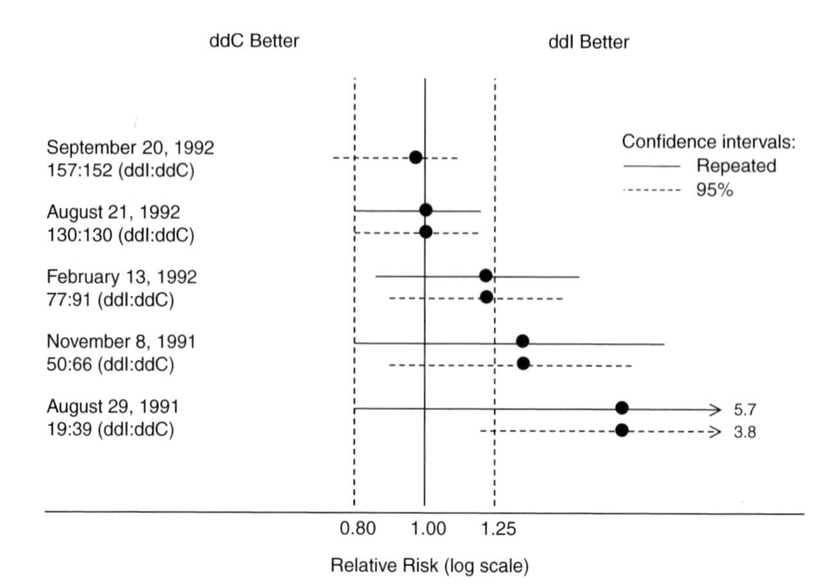

Figure 1.1 Relative risk of progression of disease (including death) by date of DMC review. Numbers to the right of the arrows are upper confidence limits. From Fleming *et al.*, Insights from monitoring the CPCRA ddI/ddC trial (1995), *Journal of Acquired Deficiency Syndromes and Human Retrovirology* **10** (Suppl. 2) Reproduced by permission of Lippincott, Williams & Wilkins.

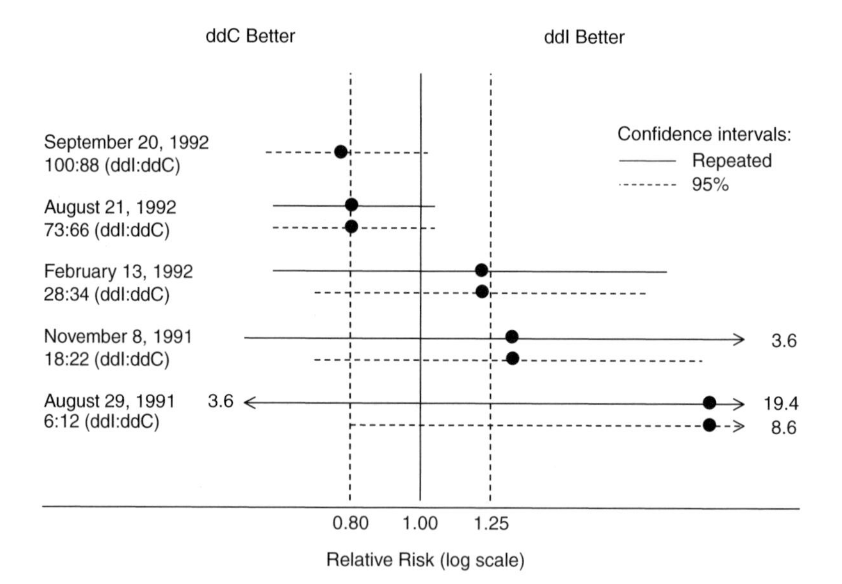

Figure 1.2 Relative risk of death by date of DMC review. Numbers to the right of the arrows are upper confidence limits. From Fleming *et al.*, Insights from monitoring the CPCRA ddI/ddC trial (1995), *Journal of Acquired Deficiency Syndromes and Human Retrovirology* **10** (Suppl. 2) Reproduced by permission of Lippincott, Williams & Wilkins.

the emerging positive data from other trials about the efficacy of ddI. Even without broad dissemination, if the data had been available to trial investigators and/or the participating pharmaceutical companies, it might have been difficult or impossible to continue the trial, given the intense pressures of the time. The investigators might have been unwilling to continue treating patients with an apparently inferior therapy; the pharmaceutical company whose product appeared superior might have chosen to end its participation and submit the available data to the Food and Drug Administration (FDA). Stopping the study early on, with a conclusion of an apparently large benefit of ddI, would clearly have been unfortunate; it would have misled patients regarding the relative efficacy of these two agents, and it would have precluded the obtaining of additional information that would ultimately contribute to the optimal continuing development of both agents as components of AIDS treatment programs.

1.2 HISTORY OF DATA MONITORING COMMITTEES IN GOVERNMENT-SPONSORED TRIALS

The concept of DMCs arose soon after the era of the modern randomized clinical trial began in the 1950s. Perhaps the first step in formalizing the concept of committees who would be charged with regular assessment of a trial's accumulating results was taken by the US National Institutes of Health (NIH). In the mid-1960s, the NIH was beginning to sponsor large, multicenter trials of new treatment interventions for serious diseases. At this time, a task force under the leadership of Dr. Bernard Greenberg of the University of North Carolina was constituted by the then National Heart Institute to develop an advisory document concerning the organization and conduct of such trials. This report, issued in 1967 (but not formally published until 1988), included among its recommendations the need for an advisory group of experts not directly involved in the conduct of the trial to review the study protocol and advise the Institute about the conduct of the trial (Heart Special Project Committee, 1988). In addition, the report addressed the need for a mechanism for terminating a trial early if it became evident that it could not meet its objectives or new information rendered it superfluous.

The influence of the 'Greenberg Report', as it came to be called, can be seen in an early trial sponsored by the NIH, the Coronary Drug Project (CDP); (CDP Research Group, 1973). This trial was initiated in the mid-1960s, and had an external committee charged with reviewing the trial conduct and the interim results on an ongoing basis. The experience in this trial reflected both the complexity of the data monitoring process and the value of an independent committee, and stimulated methodological development of new monitoring approaches.

Example 1.2: The Coronary Drug Project

The CDP was among the first, possibly the first, multicenter trial to be based on the operational model put forward by the Greenberg Report. The CDP was a multicenter, multiarm placebo-controlled trial designed to evaluate the effectiveness of five lipid-lowering treatments in patients who had experienced a cardiovascular event. More than 8000 patients were enrolled, with a planned minimum follow-up time of 5 years. A Policy Advisory Board (PAB) was initially established to review overall progress and conduct of the trial. Later, but still early in the trial, a subgroup of the PAB was formed to monitor efficacy and patient safety. This committee, which would today be called a DMC, would make recommendations to both the independent PAB as well as to the National Heart Institute. During the course of the trial, the committee recommended early termination of three of the five active treatment arms (high- and low-dose estrogen and dextrothyroxine).

The CDP Research Group (1981) published details of the consideration of interim results and the resultant decision-making. Two important themes emerged from their description of the process. First, early data trends can be very unstable, with much waxing and waning of risk ratios due to the small number of events at the early stages of the study. Thus, great caution is needed when interpreting results from early analyses. The CDP applied statistical procedures to take account of the repeated testing problem noted earlier, and they appear to be the first to describe such use in the context of a real clinical trial. These analytical approaches helped them resist the emotional pull of the early results.

The second theme was the complexity of the decision-making process, requiring multiple factors (many of which may not be readily quantifiable) to be taken into account. They noted: 'Although a number of rather sophisticated statistical tools are available in the decision making process, these are at best red flags that warn of possible treatment problems and can never be used by themselves as hard and fast decision rules' (CDP Research Group, 1981). The types of factors that need to be considered are listed in Table 1.1 and have been addressed by many DMCs since then. The continued development of statistical techniques for monitoring, far beyond what was available at the time of the CDP (see Chapter 8), has not altered the fact that statistical assessment is but one part of a highly complex decision process.

The value of DMCs to the clinical trials process was evident in the CDP, and such committees came to be a standard component of large multicenter trials sponsored by federal agencies such as the NIH and the Veterans Administration (VA). Soon after the CDP began, the National Heart Institute implemented several other trials, all with the same basic clinical trial organizational structure as the CDP, including use of a DMC. In 1968, the Urokinase Pulmonary Embolism Trial (UPET) was initiated to test the effectiveness of a thrombolytic therapy, urokinase, in resolving blood clots in the lung (UPET Study Group, 1970). This trial was followed immediately by the Urokinase-Streptokinase Pulmonary Embolism Trial

Table 1.1 Relevant factors in interim decision-making

1. Recruitment rate and completion schedule
2. Baseline characteristics and risk profile of participants
3. Baseline comparability across treatment arms
4. Compliance to intervention
5. Data completeness and follow-up
6. Internal consistency
   (a) Primary and secondary outcomes
   (b) Subgroups
   (c) Safety profile
7. External consistency
8. Statistical issues for interim analyses
9. Ethical issues
10. Impact of early termination

From DeMets DL: Methodological issues in AIDS clinical trials. Data monitoring and sequential analysis – an academic perspective (1990). *Journal of Acquired Immune Deficiency Syndrome* **3** (Suppl. 2): 5124–5133. Reproduced by permission of Lippincott, Williams & Wilkins.

(USPET); see USPET Study Group (1974). Each of these trials used a DMC. In the area of heart disease, the Hypertension Detection and Follow-up Program (HDFP) was initiated in 1972 to evaluate the impact of blood pressure reduction on 5-year mortality in individuals with mild hypertension (HDFP Cooperative Group, 1979). The Coronary Artery Surgery Study (CASS) began in 1973 to assess the effect of coronary artery bypass graft surgery compared to best medical treatment, again with 5-year mortality as the primary outcome of interest (CASS Principal Investigators, 1983). Trials in lung diseases also followed a similar model. In 1973, the Extracorporeal Membrane Oxygenator (ECMO) trial was implemented to compare a mechanical blood oxygenation device to best standard care in patients who had suffered severe lung trauma (Zapol *et al.*, 1979). The Nocturnal Oxygen Therapy Trial (1980), the Intermittent Positive Pressure Breathing Trial (1983) and the Respiratory Distress Syndrome Trial (Collaborative Group on Antenatal Steroid Therapy, 1981) were all begun in 1975. All of these early lung trials included a DMC in their organizational structures. Thus, by the mid-1970s, the Institute (by then renamed the National Heart, Lung and Blood Institute (NHLBI)) was routinely establishing DMCs to monitor randomized clinical trials in all clinical areas.

In 1972, two senior statisticians left the NHLBI to establish a biometrics research group in the newly formed National Eye Institute (NEI). They brought with them the knowledge and experience of NHLBI's emerging clinical trials programs, including the role of the DMC. This influence can be seen in the Diabetic Retinopathy Study (DRS), one of the first NEI randomized trials, which evaluated a new photocoagulation treatment in diabetic patients experiencing proliferative retinopathy (DRS Research Group, 1976) by randomly assigning one eye of each study participant to the new treatment, and the other eye to standard management. The DMC for this study made a significant protocol change early in the trial, based on an unexpected early large benefit of the photocoagulation treatment. Rather than recommend early termination, the DMC recommended that each 'control eye' should receive the photocoagulation treatment when it reached a specified level of retinopathy. This change permitted the evaluation of safety and duration of benefit based on longer follow-up. The NEI has routinely incorporated DMCs into the structure of the major Phase III comparative trials it sponsors.

In the mid-1970s, the VA first developed guidelines for their Cooperative Group network for conducting clinical trials for VA patients. These guidelines, which are regularly revised and updated, include the use of DMCs (Cooperative Studies Program, 2001).

Cancer trials sponsored by the National Cancer Institute (NCI) began to use DMCs in the early 1980s. The first group to establish DMCs was the North Central Cancer Treatment Group (NCCTG), headquartered at the Mayo Clinic. This group developed a DMC model that, instead of outside experts, involved study investigators and a statistician from the group's coordinating center. Although this group was clearly not independent of the trials, it established the key approach of not sharing interim data widely among all investigators, as was the common practice in cancer trials at that time; access to the interim data was limited to the DMC. Soon afterwards (and facilitated by a move of two NCCTG statisticians involved in that DMC to the statistical center for the Southwestern Oncology Group (SWOG)), the concept of the DMC was introduced to SWOG and adopted as part of SWOG operating procedures. Shortly thereafter an 'intergroup' study of adjuvant therapy of colon cancer (#0035) was initiated with clinical leadership from the NCCTG and statistical leadership from SWOG. The trial incorporated the DMC model developed by the NCCTG, and provided the first opportunity for the other cooperative groups to experience this approach. Other cancer cooperative groups established DMCs following issuance of data monitoring policies by the NCI in 1994 (Smith *et al.*, 1997). NCI-sponsored cancer prevention trials such as the Alpha-tocopherol, Beta-Carotene (ATBC) lung cancer prevention trial in Finland (ATBC Cancer Prevention Study Group, 1994) and the Beta-Carotene and Retinol Efficacy Trial (CARET) study (Omenn *et al.*, 1996) in the United States also had formal DMCs. It is interesting to note that despite the fact that cancer treatment trials (unlike most trials in cardiovascular disease) are generally unblinded because of the complex nature of administration of chemotherapy and the distinctive toxicities associated with different agents, DMCs have been found to be a valuable component of clinical cancer research.

The AIDS epidemic that emerged in the early 1980s led NIAID to form two clinical trial networks, the AIDS Clinical Trial Group (ACTG) and the CPCRA. These two NIAID-sponsored clinical trials groups were served by a single DMC that had to develop new operational approaches to deal with the many new challenges posed by HIV/AIDS trials (DeMets *et al.*, 1995; Ellenberg *et al.*, 1993b). These challenges included having to monitor multiple trials from two different organizational groups, an unprecedented involvement of patient representatives and advocacy groups in the design of trials and interpretation of results, and scientific and political pressures to identify effective treatments as quickly as possible. In addition, the pharmaceutical industry was much more closely involved with these trials than had been traditional with NIH clinical trials programs, requiring careful attention to lines of communication with industry sponsors.

Clinical trials in Europe have also used DMCs for trial oversight. The International Studies of Infarct Survival (ISIS), initiated in the 1970s, have always incorporated DMCs in their trial structures. The Medical Research Council in the UK instituted a policy of establishing independent DMCs for 'high-profile' trials in the early 1990s (Parmar and Machin, 1993). In other countries, as in the USA, DMCs were first seen primarily in cardiovascular trials (19).

1.3 DATA MONITORING COMMITTEES IN TRIALS SPONSORED BY THE PHARMACEUTICAL INDUSTRY

While DMCs were widely used in government-sponsored trials, DMCs were only occasionally established for trials sponsored by the pharmaceutical industry until the early 1990s. Of the few industry-sponsored trials that did establish formal DMCs, many were in the area of cardiovascular disease with improvement of survival as the primary goal (Anturane Reinfarction Trial Research Group, 1980; Persantine-Aspirin Reinfarction Study Research Group, 1980; Swedberg *et al*., 1992; APSAC Intervention Mortality Study Trial Group, 1988; European Myocardial Infarction Project, 1988), following the model established by the NHLBI. The move to greater use of DMCs in industry trials may have been influenced at least in part by three factors. First, concerns emerging in the late 1980s and early 1990s about the reliability of surrogate endpoints led to increased numbers of industry trials designed to directly assess effects on clinical endpoints such as mortality. Second, the increased collaborative efforts of industry and NIH in areas such as cardiovascular and AIDS research exposed pharmaceutical companies to clinical trial models that were new to them, particularly with regard to the assignment of responsibility for interim monitoring to an independent committee. These activities brought them into increased contact with researchers whose experience with DMCs led them to strongly advocate their use in industry trials. Third, although DMCs are not generally required for clinical trials performed by regulated industry (see Chapter 10), FDA staff increasingly recommend that companies establish DMCs for certain types of trials.

While many companies had concerns about giving up access to the accumulating data as the trial progressed, the establishment of DMCs offered companies some clear advantages. For example, regulatory bodies traditionally have been uncomfortable about companies making changes to trials in progress, recognizing the potential biases that can arise in making such decisions because of the large financial stake a company has in the outcome of its trials. Even when the rationale for such changes appears well founded, one cannot ignore the concern that a company might identify and recommend only those changes that are likely to be advantageous to the company. When changes are proposed by sponsors who do not have access to interim trial data, concerns about bias are substantially diminished. In addition, independent DMCs may protect companies against claims of misleading stockholders, as in the next example.

Example 1.3: Treatment of amyotrophic lateral sclerosis

A multicenter randomized double-blind placebo-controlled trial, the ALS CNTF Treatment Study (ACTS), evaluated a new ciliary neurotrophic factor (CNTF) in patients with amyotrophic lateral sclerosis (ALS) or Lou Gehrig's disease. In ALS, the patient's muscle strength, including respiratory function, rapidly deteriorates. Survival time from onset of disease is typically only a few years. The new nerve growth factor was believed to increase muscle strength or at least prevent further deterioration. ACTS was sponsored by a small biotech company that established an independent steering committee, an independent DMC and an independent statistical center for interim analyses. Thus, the sponsor was totally blinded during the conduct of the trial. The DMC terminated the trial early due to adverse effects, observing that measures of muscle strength were worse on the new therapy than on the placebo. Results of the trial have been published (ALS CNTF Treatment Study Group, 1996). Following the decision to terminate, the DMC immediately briefed the steering committee and the sponsor. The sponsor, within a day, alerted the financial community. Later, investors who had had great expectations for this new therapy brought legal action against the sponsor, arguing that the sponsor had misled them by not alerting them earlier about the impending negative results (Wall Street Journal, 1994). Since the sponsor was kept blinded to accumulating results during the course of the trial, they did not know the results until the day before the results were made public. The use of an independent DMC provided the sponsor with a strong defense against such claims of illegal activity.

As the use of DMCs in clinical trials increased, the existence of widely varying policies and practices for such committees by trial sponsors became evident. At an international conference held at the NIH in 1992, many differing views on the optimal approaches to data monitoring were presented (Ellenberg *et al.*, 1993a). In recent years, papers describing and/or advocating specific monitoring practices have appeared in greater numbers (Armitage, 1999a, 1999b; Armstrong and Furberg, 1995; Canner, 1983; DeMets *et al.*, 1982, 1984, 1995, 1999; Dixon and Lagakos, 2000; Fleming and DeMets, 1993; Freidlin *et al.*, 1999; Meinert, 1998a, 1998b; Pocock, 1993; Whitehead, 1999). Another recent development has been mention of DMCs for the first time in regulations and guidance documents of regulatory authorities such as the FDA (Code of Federal Regulations, Title 21, Part 50.24; US Food and Drug Administration, 1997, 1998, 2001).

1.4 STATISTICAL METHODS FOR INTERIM MONITORING

The practical experiences and challenges faced by DMC members in the 1970s led to the development of statistical methods that accounted for the multiplicity problem generated by repeated conduct of interim analyses at scheduled DMC meetings. These approaches, known as 'group sequential methods' to distinguish them from earlier approaches based on assumptions of continual interim analysis (Pocock, 1977; O'Brien and Fleming, 1979; Lan and DeMets, 1983), were rapidly adopted during the 1980s and provided a new structure to the data monitoring process.

The fundamental statistical problem is fairly simple. Under the null hypothesis in an intrinsically one-sided superiority or non-inferiority setting, one wishes to maintain an upper bound on the false positive error rate, at a nominal level that usually is 2.5%. (The standard for strength of evidence, corresponding to allowing a 2.5% false positive error rate, is achieved whether one is conducting a two-sided 0.05-level test or a one-sided 0.025-level test.) However, the performance of multiple tests of the null hypothesis over time will lead to a false positive error rate substantially higher than the nominal level at which the tests are performed. Thus, if we test our data frequently during the course of a trial, the probability that we will, at some point, observe a difference that is 'statistically significant' at the one-sided 0.025 level is in fact substantially greater than 2.5%. To preserve the desired level of false positive error, it is necessary to perform interim testing at more conservative levels.

Several approaches to this problem can be taken. Perhaps most simply, one can perform all interim tests at highly conservative levels (e.g., require a p-value of 0.001 or less to justify early termination with a conclusion of strong evidence against the null hypothesis), so that the impact on overall false positive error is minimal and final testing can be done at the conventional level without much worry about having inflated the false positive rate. This approach, first proposed by Haybittle (1971), is very conservative, even when the trial nears the time of completion. Another straightforward approach, described by Pocock (1977), uses the same significance level for all interim tests as well as the final test, with this level calculated to provide the desired overall false positive error. In order to calculate the testing levels to be used, the number of interim tests must be specified. At interim analyses, this approach is much less conservative than the first approach; unless a huge number of interim tests is specified, the required significance level for all tests will be substantially less stringent than 0.001. One difficulty with the Pocock approach is that, unlike the Haybittle approach, it requires the final test be performed at a lower than conventional level. For example, if four interim analyses are planned and an overall 0.025 false positive error rate is desired, the final (fifth) analysis will need to be done at a one-sided significance level of approximately 0.008. This allows for the uncomfortable situation of observing a final difference that produces a one-sided p-value of 0.01 but being unable to reject the null hypothesis at the one-sided 0.025 level.

To achieve the intuitively appealing property of the Haybittle approach (permitting final testing at nearly the conventional significance level) without its extreme conservatism in the latter stages of trial monitoring, O'Brien and Fleming (1979) developed an alternative to Pocock's approach that varied the significance level used for interim testing as the trial progressed. Their method provides for highly conservative criteria early in the trial, with progressively less stringent criteria at successive interim analyses, and a final analysis that can be performed at close to the nominal level.

The O'Brien–Fleming group sequential boundary is one member of a family of boundaries with similar characteristics that can be generated (Wang and Tsiatis, 1987). In this book, most of the examples presented use this statistical approach. The O'Brien–Fleming type boundary has become popular, probably because its properties reflect the thinking of many of those with experience in evaluating interim trial results. First, this boundary is very conservative early in the trial when the numbers of patients and events are small and any estimate of treatment effect is therefore unreliable, requiring great caution in interpretation. Second, as more patients are recruited and more events are observed, the information fraction increases and the O'Brien–Fleming criteria for statistical significance become correspondingly less stringent. A third reason is that at the completion of the trial (assuming termination did not occur earlier) the critical value for the test statistic is close to the nominal value (e.g., 0.05, two-sided), the same critical value that would be used for a trial with no sequential testing. Consequently, the power of the trial is maintained without having to increase the sample size. This is important because it permits the conduct of interim analyses while maintaining the false positive error rate at accepted levels without having to substantially increase trial size and cost. These and other methods will be discussed in more detail in Chapter 8.

1.5 WHEN ARE DATA MONITORING COMMITTEES NEEDED?

As noted at the beginning of this chapter, DMCs are most relevant to randomized clinical trials specifically focused on clinical efficacy and safety. They have been used primarily for trials that are expected to provide a definitive answer to a question about whether a drug is effective, or whether one drug regimen is more effective than another. Further, even in this setting they have been used mostly in trials that address major health outcomes such as mortality, progression of a serious disease, or occurrence of a life-threatening event such as heart attack and stroke. They have not been used as widely in the many randomized trials (nearly always short-term) that address symptom relief, nor in trials implemented early in drug development whose results will be examined in an exploratory fashion and whose successor trials will be looked to for definitive conclusions.

Although the monitoring of trials with regard to safeguarding the interests of study participants and to ensuring that the trial is being conducted properly is appropriate and necessary in every clinical study, a formal DMC is not routinely needed. We propose several general criteria that can help determine the need for and value of a DMC in a given trial:

1. Is the trial intended to provide definitive information about effectiveness and/or safety of a medical intervention?
2. Are there prior data to suggest that the intervention being studied has the potential to induce potentially unacceptable toxicity?
3. Is the trial evaluating mortality or another major endpoint, such that inferiority of one treatment arm has safety as well as effectiveness implications?
4. Would it be ethically important for the trial to stop early if the primary question addressed has been definitively answered, even if secondary questions or complete safety information were not yet fully addressed?

A DMC usually should be implemented if two or more of these criteria are met, and usually would not be considered if none are met. In some cases, when the treatments are novel and raise serious safety questions, DMCs are used even in early, non-randomized studies – such studies would meet only criterion 2. Considerations for use of DMCs will be further addressed in Chapter 9.

1.6 WHERE WE ARE TODAY

At present, DMCs appear to be a 'growth industry'. The incorporation of such committees into trial structures is increasing, not only in industry trials, but also in government-sponsored trials in disease areas (such as cancer) that did not initially use DMCs. With this rapid growth, a variety of approaches to DMC operations have been developed for trials in multiple medical areas. Although there is broad agreement on many principles and procedures relating to DMC functions, there are important aspects for which consensus currently does not exist. Further, as with any area where ethical issues arise, opinions tend to be strongly held among knowledgeable and experienced individuals regarding the optimal approach to interim monitoring and the operation of DMCs.

In this book, we will present principles and guidelines for constituting and implementing DMCs and for establishing the policies and procedures under which they operate. We will describe the variety of approaches that have been taken to address the operational aspects of the data monitoring of clinical trials, and will give special attention to some of the controversies that have arisen regarding optimal practices. While attempting to lay out the advantages and disadvantages of the various approaches, we will not hesitate to recommend, based on our own experience in a wide range of medical settings, what we believe to be the best way to proceed.

1.7 FUNDAMENTAL PRINCIPLES OF DATA MONITORING

We conclude this Introduction with a brief discussion of some fundamental principles that will be invoked throughout the book, and that will be addressed in more detail in later chapters.

Principle 1. The primary responsibilities of a DMC are to: (i) safeguard the interests of study patients; (ii) to preserve the integrity and credibility of the trial in order that future patients may be treated optimally; and (iii) to ensure that definitive and reliable results be available in a timely way to the medical community.

The DMC has responsibilities to the trial investigators and sponsor, and to the scientific community generally, to ensure the integrity and reliability of the scientific result that is obtained. As discussed more fully in Chapter 2, however, its primary responsibility is to ensure the safety of the study participants. It must provide assurance to patients and their treating physicians that patients' care will not be compromised because of participation in the study.

Principle 2. The DMC should have multidisciplinary representation, including physicians from relevant medical specialties and biostatisticians. In many cases, other experts such as bioethicists, epidemiologists and basic scientists should also be included.

Due to the complexity of clinical trials and the decision-making process (discussed in detail in Chapter 3), the DMC requires sufficiently broad membership to ensure that all relevant medical, ethical, safety and scientific issues can be adequately discussed and properly weighed in all recommendations concerning trial conduct and termination.

Principle 3. The DMC should have membership limited to individuals free of apparent significant conflicts of interest, whether they are financial, intellectual, professional or regulatory in nature.

There is an intrinsic need for judgment in the process of developing recommendations about important study conduct issues, including whether to terminate or continue a trial. Such recommendations must be perceived to have been made in a fair and unbiased manner. Study integrity and credibility are compromised if individuals with apparent conflicts of interest influence recommendations. This could occur, for example, if members of a DMC were in a position to realize financial or professional gain should the study produce a positive result. Because of this concern, discussed more fully in Chapter 4, sponsors or others having significant financial or professional interests that could be affected by the outcome of the trial should generally not be members of the DMC for that trial. The word 'significant' is key; it may be necessary in some cases to include individuals with some potential conflict of interest if there are no other individuals with the requisite expertise available to serve on the DMC. In such cases, disclosure of the potential conflict – to other committee members, the sponsor, the regulatory agency, etc. – is necessary. The disclosure process permits independent assessments about whether the apparent conflict might have had a significant impact.

Principle 4. The DMC members should ideally be the only individuals to whom the data analysis center provides results on relative efficacy and safety of study treatments.

As we saw in the CPCRA 002 example, it is common for early results to be misleading by giving the inaccurate impression that treatment effects are markedly favorable or unfavorable. Thus, as will be discussed in detail in Chapter 5, widespread reporting of interim results greatly increases the risk of actions being taken based on unreliable information. Such actions could include inappropriate early abandonment of the trial, or even of other related trials. In the setting of oncology trials, Green *et al.* (1987) demonstrated that providing a DMC sole access to the interim data reduced the risk that trials would experience declining accrual rates over time, inappropriate early termination yielding equivocal results, or final results that were inconsistent with prematurely published early results (see also Armitage, 1999b). If it appears necessary in specific cases to make interim efficacy and safety results available to individuals outside the DMC, those individuals should agree to maintain the confidentiality of the information.

REFERENCES

Abrams D, Goldman A, Launer C *et al.* (1994) A comparative trial of didanosine or zalcitabine after treatment with zidovudine in patients with human immunodeficiency virus infection. *New England Journal of Medicine* **330**: 657–662.

Alpha-Tocopherol, Beta-Carotene Cancer Prevention Study Group (1994) The effect of vitamin E and beta carotene on the incidence of lung cancer and other cancers in male smokers. *New England Journal of Medicine* **330**: 1029–1035.

ALS CNTF Treatment Study Group (1996) A double-blind placebo-controlled clinical trial of subcutaneous recombinant human ciliary neurotrophic factor (rHCNTF) in amyotrophic lateral sclerosis. *Neurology* **46**: 1244–1249.

Anturane Reinfarction Trial Research Group (1980) Sulfinpyrazone in the prevention of sudden death after myocardial infarction. *New England Journal of Medicine* **302**: 250–256.

APSAC Intervention Mortality Study Trial Group (1988) Effect of intravenous APSAC on mortality after acute myocardial infarction: preliminary report of a placebo-controlled clinical trial.*Lancet* **1**: 545–549.

Armitage P on behalf of the Concorde and Alpha Data and Safety Monitoring Committee (1999a) Data and safety monitoring in the Concorde and Alpha Trials. *Controlled Clinical Trials* **20**: 207–228.

Armitage P on behalf of the Delta Data and Safety Monitoring Committee (1999b) Data and safety monitoring in the Delta Trial. *Controlled Clinical Trials* **20**: 229–241.

Armstrong PW, Furberg CD (1995) Clinical trials and safety monitoring boards: the search for a constitution. *Circulation* **91**: 901–904.

Buyse M (1993) Interim analyses, stopping rules and data monitoring in clinical trials in Europe. *Statistics in Medicine* **12**: 509–520.
Canner PL (1983) Monitoring of the data for evidence of adverse or beneficial treatment effects. *Controlled Clinical Trials* **4**: 467–483.
CASS Principal Investigators & Their Associates (1983) Coronary Artery Surgery Study (CASS): A randomized trial of coronary artery bypass surgery. *Circulation* **68**(5): 939–950.
Collaborative Group on Antenatal Steroid Therapy (1981) Effect of antenatal dexamethasone administration on the prevention of respiratory distress syndrome. *American Journal of Obstetrics and Gynecology* **141**: 276–287.
Cooperative Studies Program (2001) *Guidelines for the Planning and Conduct of Cooperative Studies*. Office of Research and Development, Department of Veterans Affairs. http://www.va.org/resdev.
Coronary Drug Project Research Group (1973) The Coronary Drug Project: design, methods, and baseline results. *Circulation* **47** (Suppl. 1), xlvii–xlviii.
Coronary Drug Project Research Group (1981) Practical aspects of decision-making in clinical trials: the Coronary Drug Project as a case study. *Controlled Clinical Trials* **1**: 363–376.
DeMets DL, Williams GW, Brown Jr. BW and the NOTT Research Group (1982) A case report of data monitoring experience: the Nocturnal Oxygen Therapy Trial. *Controlled Clinical Trials* **3**: 113–124.
DeMets DL, Hardy R, Friedman LM, Lan KKG (1984) Statistical aspects of early termination in the Beta-Blocker Heart Attack Trial. *Controlled Clinical Trials* **5**: 362–372.
DeMets DL, Fleming TR, Whitley RJ, Childress JF, Ellenberg SS, Foulkes M, Mayer KH, O'Fallon J, Pollard RB, Rahal JJ, Sande M, Straus S, Walters L, Whitley-Williams P (1995) The Data and Safety Monitoring Board and acquired immune deficiency syndrome (AIDS) clinical trials. *Controlled Clinical Trials* **16**: 408–421.
DeMets DL, Pocock SJ, Julian DG (1999) The agonising negative trend in monitoring of clinical trials. *Lancet* **354**: 1983–1988.
Diabetic Retinopathy Study Research Group (1976) Preliminary report on effects of photocoagulation therapy. *American Journal of Ophthalmology* **81**: **383**.
Dixon DO, Lagakos SW (2000) Should data and safety monitoring boards share confidential interim data? *Controlled Clinical Trials* **21**: 1–6.
Ellenberg SS, Geller N, Simon R, Yusuf S (eds) (1993a) Proceedings of 'Practical issues in data monitoring of clinical trials', Bethesda, Maryland, USA, 27–28 January 1992. *Statistics in Medicine* **12**: 415–616.
Ellenberg SS, Myers MW, Blackwelder WC, Hoth DF (1993b) The use of external monitoring committees in clinical trials of the National Institute of Allergy and Infectious Diseases. *Statistics in Medicine* **12**: 461–467.
European Myocardial Infarction Project (1988) Potential time saving with pre-hospital intervention in acute myocardial infarction. *European Heart Journal* **9**, 118–124.
Fleming TR, DeMets DL (1993) Monitoring of clinical trials: issues and recommendations. *Controlled Clinical Trials* **14**:183–197.
Freidlin B, Korn EL, George SL (1999) Data monitoring committees and interim monitoring guidelines. *Controlled Clinical Trials* **20**: 395–407.
Green SJ, Fleming TR, O'Fallon JR (1987) Policies for study monitoring and interim reporting of results. *Journal of Clinical Oncology* **5**: 1477–1484.
Haybittle JL (1971) Repeated assessment of results in clinical trials of cancer treatments. *British Journal of Radiology* **44**: 793–797.
Heart Special Project Committee (1988) Organization, review and administration of cooperative studies: A report from the Heart Special Project Committee to the National Advisory Heart Council, May 1967. *Controlled Clinical Trials* **9**: 137–148.

Hypertension Detection and Follow-up Program Cooperative Group (1979) Five-year findings of the hypertension detection and follow-up program. I. Reduction in mortality of persons with high blood pressure, including mild hypertension. *Journal of the American Medical Association* **242**(23): 2562–2571.

Intermittent Positive Pressure Breathing Trial Group (1983) Intermittent positive pressure breathing therapy of chronic obstructive pulmonary disease. *Annals of Internal Medicine* **99**(5): 612–620.

Lan KKG, DeMets DL (1983) Discrete sequential boundaries for clinical trials. *Biometrika* **70**: 659–663.

Meinert CL (1998a) Clinical trials and treatment effects monitoring. *Controlled Clinical Trials* **19**: 515–521.

Meinert CL (1998b) Masked monitoring in clinical trials – blind stupidity? *New England Journal of Medicine* **338**: 1381–1382.

Nocturnal Oxygen Therapy Trial Group (1980) Continuous or nocturnal oxygen therapy in hypoxemic chronic obstructive lung disease. A clinical trial. *Annals of Internal Medicine* **93**(3): 91–98.

O'Brien PC, Fleming TR (1979) A multiple testing procedure for clinical trials. *Biometrics* **35**: 549–556.

Omenn GS, Goodman GE, Thornquist MD, Balmes J, Cullen MR, Glass A, Keogh JP, Meyskens FL Jr, Valanis B, Williams JH Jr, Barnhart S, Hammar S (1996) Effects of a combination of beta carotene and vitamin A on lung cancer and cardiovascular disease. *New England Journal of Medicine* **334**: 1150–1155.

Parmar MKB, Machin D (1993) Monitoring clinical trials: experience of, and proposals under consideration by, the cancer therapy committee of the British Medical Research Council. *Statistics in Medicine* **12**: 497–504.

Persantine-Aspirin Reinfarction Study Research Group (1980) Persantine and aspirin in coronary heart disease. *Circulation* **62**: 449–461.

Pocock SJ (1977) Group sequential methods in the design and analysis of clinical trials. *Biometrika* **64**: 191–199.

Pocock SJ (1993) Statistical and ethical issues in monitoring clinical trials. *Statistics in Medicine* **12**: 1459–1469.

Smith MA, Ungerleider RS, Korn EL, Rubinstein L, Simon R (1997) Role of independent data-monitoring committees in randomized clinical trials sponsored by the National Cancer Institute. *Journal of Clinical Oncology* **15**: 2736–2743.

Swedberg K, Held P, Kjekshus J, Rasmussen K, Ryden L, Wedel H (1992) Effects of early administration of enalapril on mortality in patients with acute myocardial infarction: results of the Cooperative North Scandinavian Enalapril Survival Study II. *New England Journal of Medicine* **327**: 678–684.

Urokinase Pulmonary Embolism Trial Study Group (1970) Urokinase pulmonary embolism trial: phase I results. A cooperative study. *Journal of the American Medical Association* **214**(12): 2163–2172.

Urokinase-Streptokinase Pulmonary Embolism Trial Study Group (1974) Urokinase-streptokinase pulmonary embolism trial: phase II results. A cooperative study. *Journal of the American Medical Association* **229**(12): 1606–1613.

US Food and Drug Administration (1997) Guidance for industry: Good clinical practices. *Federal Register* **62**: 25 691–25 709.

US Food and Drug Administration (1998) Guidance for industry: Statistical principles in clinical trials. *Federal Register* **63**: 49 583–49 598.

US Food and Drug and Administration (2001) *Guidance for Clinical Trial Sponsors on the Establishment and Operation of Clinical Trial Data Monitoring Committees*. Rockville, MD: FDA. http://www.fda.gov/cber/gdlns/clindatmon.htm.

Wall Street Journal (1994) Small fast-growth firms feel chill of shareholder suits. *Wall Street Journal*, April 5.

Wang SK, Tsiatis AA (1987) Approximately optimal one-parameter boundaries for group sequential trials. *Biometrics* **39**: 193–199.
Whitehead J (1999) On being the statistician on a data and safety monitoring board. *Statistics in Medicine* **18**: 3425–3434.
Zapol WM, Snider MT, Hill JD *et al.* (1979) Extracorporeal membrane oxygenation in severe acute respiratory failure. A randomized prospective study. *Journal of the American Medical Association* **242**(20): 2193–2196.

2

Responsibilities of the Data Monitoring Committee and Motivating Illustrations

Key Points

- The DMC may perform a variety of tasks in meeting its overall responsibilities for safeguarding trial participants and trial integrity.
- These tasks may address primarily scientific issues, such as an advisory review of the study protocol, as well as primarily practical issues such as quality assurance.
- Trials may be considered for early termination based on proven efficacy, unfavorable results ruling out benefit, safety concerns, or low probability of achieving the trial objectives.
- All trials monitored by a DMC should function under a charter, agreed to by the trial sponsor and the committee members, that describes the structure and operation of the committee.

2.1 FUNDAMENTAL CHARGES

In randomized trials designed to evaluate the relative efficacy and safety of interventions, particularly in the setting of life-threatening diseases, it is important to monitor evolving information regarding the benefits and risks of these interventions. Such review satisfies important ethical needs to protect the interests of study participants and to provide timely insights to the broader clinical community. These periodic interim reviews, together with other types of trial monitoring by the sponsor and investigators, also enable the collection of higher-quality information by allowing for timely modifications of procedures for patient accrual and management and for data collection.

The fundamental charges to those responsible for trial monitoring, as discussed in Chapter 1, should have the following prioritization: first, to safeguard the interests of the study participants; second, to preserve the trial integrity and credibility; and third, to facilitate the availability of timely as well as reliable findings to the broader clinical community. An important element of these fundamental charges is the need to determine the ethical and scientific appropriateness of continuing the clinical trial. Periodic review of evolving data allows termination of the trial if early results about the benefit-to-risk profile of the experimental therapy are convincingly positive or negative. Statistical procedures, such as group sequential monitoring boundaries, provide useful insights regarding the strength of evidence required to justify a recommendation for termination. However, because of the complexity of randomized clinical trials, these statistical procedures are intended to provide helpful guidelines, rather than rigid rules, about whether early trial termination should occur. Recommendations about trial termination or continuation must be based on a global consideration of all available data from the trial, including information on primary and secondary efficacy measures, adverse effects, and quality of trial conduct, along with relevant information external to the trial.

Necessarily, then, well-informed and scientifically objective judgments are required to integrate this global information and arrive at these recommendations. A data monitoring committee can provide an appropriate structure through which these well-informed and scientifically objective judgments can be made (Coronary Drug Project Research Group, 1981; Heart Special Project Committee, 1988; Fleming and DeMets, 1993; DeMets *et al.*, 1995). Some fundamental principles, again discussed in Chapter 1, should be considered in defining the composition and functioning of these committees. To be well informed (see Chapter 3), the DMC should have multidisciplinary representation, including physicians from relevant medical disciplines, biostatisticians, and often ethicists or other experts, preferably all with experience in and understanding of clinical trials. To provide objective judgments (see Chapter 4), the DMC should have membership limited to individuals free of apparent significant conflicts of interest, whether financial, professional or regulatory in nature. To reduce the risk of widespread prejudgment of unreliable results based on limited data (see Chapter 5), the DMC should ideally be the only individuals to whom the data analysis center provides interim results on relative efficacy and safety of treatment regimens.

In most instances, actions by the DMC to address its primary responsibilities will also serve the interests of the study investigators and the government or industry sponsor. However, conflicts can arise. For example, the need to maintain confidentiality of study results, discussed in Chapter 5, can conflict with the desire of a sponsor to have early access to interim study data in order to support an early submission with a regulatory agency, or to make decisions about further development of a medical product in other populations, or to inform decisions regarding whether or when to ramp up production capabilities in the event the trial is positive. When conflicting needs arise, the DMC must seek approaches that will be as helpful as possible to sponsors or other parties while compromising neither the care of patients nor the trial's ability to provide conclusive evidence about the safety and efficacy of study interventions.

In its advisory capacity to the study investigators and sponsor, the DMC will make recommendations about when to terminate the clinical trial. In this capacity, the DMC may also make recommendations about modifications in study design or conduct, or in procedures for data management, quality control or reporting, as will be discussed later in this chapter. In turn, the study investigators and sponsor, who retain the ultimate responsibility for the design, conduct and reporting of the trial, should promptly review such recommendations and decide on the course of action they judge to be in the best interest of the study and its patients. A steering committee, consisting of representatives of the investigators and sponsor, can facilitate this decision-making process. The role of the steering committee will be explored in Chapter 7.

As noted by the Coronary Drug Project Research Group (1981) and by Green *et al.* (1987), judgments about whether to continue a clinical trial should consider three groups of patients: those already enrolled on study, those yet to be entered, and current and future patients outside the study. Premature termination of a trial can produce significant negative consequences for each of these groups. As Green *et al.* put it:

> The commitment and cooperation of patients currently on study are wasted if a study becomes equivocal or misleading. Thousands of future patients are at risk for receiving an ineffective or costly or toxic treatment (if a treatment is erroneously reported as superior) or are at risk for not receiving an effective treatment (if a new treatment is erroneously reported as being not better than a standard). Even patients yet to be entered on study, the very patients we most seek to protect by early termination, are not necessarily helped by such action, since they are likely to receive the regimen that appeared preferable in early data, even though more data or longer follow-up might have shown it to be inferior.

The responsibilities to all patients are best served by charging well-informed DMCs with the responsibility for making recommendations for continuation or early termination.

2.2 SPECIFIC TASKS OF THE DATA MONITORING COMMITTEE

In the previous section, a key responsibility of the DMC was indicated to be the development of recommendations about trial continuation. In this section, the specific tasks of the DMC will be discussed that enable it not only to meet this responsibility but also to address a broader role relative to enhancement of the quality of trial conduct. These tasks should be summarized in a DMC charter that provides a detailed summary of the standard operating procedures regarding the role and functioning of the DMC. The DMC charter is discussed later in this chapter, and a sample is provided in Appendix A.

As would be anticipated, the most important tasks of the DMC relate to monitoring safety and efficacy data while the trial is ongoing. However, this section begins with a discussion of potentially valuable contributions by the DMC before such data would be available.

2.2.1 Initial review

The study investigators and sponsor have the primary responsibility for the development of the study protocol and the procedures to ensure quality of study conduct. In many instances, the DMC will be asked to review these documents prior to initiation of the trial. By providing an advisory review of the draft of the protocol and proposed study procedures, the DMC can ensure that none of its members have concerns about the planned trial that would interfere with their ability to monitor the study in the manner specified by the sponsor and investigators. This initial review also allows the DMC to give independent scientific guidance and reduce the risk that ethical or scientific flaws would be identified during the course of study conduct.

2.2.1.1 Review of the study protocol

In order for the DMC to be able to carry out its primary responsibilities of safeguarding the interests of study participants and preserving the integrity and credibility of the trial, its members should be supportive of the protocol-specified study objectives and design. If the DMC members believe the study design to be flawed in important ways, it may be difficult for them to meet these responsibilities. For example, one or more members of the DMC might view some aspects of the recruitment process as unacceptably coercive, or have concerns that the informed consent procedures may be misleading, confusing or inadequately informative regarding what is already known about the risks and benefits of study interventions, or that proposed statistical monitoring procedures do not appear to adequately safeguard study participants. The DMC might also raise important scientific issues, such as concerns about whether the primary endpoint is an acceptable basis for assessing treatment benefit. Such concerns about the primary endpoint could make the DMC uncomfortable about recommending early termination, even though the interim data would suggest that the treatment effects achieve the statistical criteria for early termination.

Thus, arrangements should be made to enable the DMC to review the study protocol at the time of completion of its penultimate draft. This timing permits any DMC-recommended modifications to be considered by the sponsor and/or steering committee. In its advisory capacity to the investigators and sponsor, the DMC might provide additional recommendations for improving the design. Typically, such recommendations would be of limited scope, since the DMC's primary role is to be one of independent oversight. The investigators and sponsor would then address what modifications would be made in finalizing the protocol. By way of illustration, we will consider a clinical trial sponsored by industry and two trials sponsored by the National Institutes of Health.

Example 2.1: Gamma interferon in chronic granulomatous disease

Chronic granulomatous disease (CGD) is a rare condition arising in some children having a compromised immune system, and leads to recurrent, serious and often life-threatening infections. Gamma interferon was considered a promising treatment for this disease, due to its documented effect on the immune system. The expectation was that it would increase superoxide production and bacterial killing, effects that would be expected to reduce susceptibility to infections. As a result, an industry-sponsored placebo-controlled trial of gamma interferon was planned (International Chronic Granulomatous Disease Study Group, 1991).

Some of the study investigators proposed treating patients for 12 months in order to be able to assess the effect of treatment on the clinical endpoint of recurrent serious infections. Many other investigators and the sponsor favored a much shorter treatment duration of 1 month, since that would be sufficient to assess the impact of treatment on biological markers thought to predict the clinical outcome, would avoid the need to give three weekly injections of the placebo for a year to one-half of the children, and would permit the treatment to be made available more rapidly should the trial results be positive. The sponsor formed a DMC comprised of one statistician and three physicians who were experts in CGD but not involved in treating study patients. The sponsor also arranged for DMC members to attend the study investigators' final protocol development meeting to provide the DMC greater insight into the difficult design issues these investigators were confronting. At this meeting, the DMC made the recommendation that a 12-month study, in which the effect of the treatment on the rate of infection would be assessed, would be preferable to the shorter study focusing on the biological markers. There could, in turn, be an interim analysis guided by a group sequential design (see Chapter 8) that would allow for early termination, if early results were extreme.

The study team agreed to conduct the longer study, partially motivated by the DMC recommendation. Results after 6 months provided strong evidence that treatment substantially affected the clinical endpoint, the rate of recurrent serious infections, and the trial was terminated at that time. In contrast, however, treatment with gamma interferon had no detectable effect on either superoxide production or bacterial killing. This example illustrates not only the considerable risks of misleading conclusions when surrogate markers replace clinical efficacy endpoints, but also the constructive role a DMC might have in the final stages of study design.

Example 2.2: Peptide T in HIV-infected patients

In 1991, the National Institute of Mental Health (NIMH) formed a DMC for its controlled evaluation of peptide T in HIV-infected patients with cognitive impairment (Heseltine *et al.*, 1998). A pre-trial meeting of the DMC with the study investigators and NIMH program officers was held in the fall of 1991 to review several design issues. These included a plan for treatment crossover for each patient completing 6 months of study therapy, and the intent to exclude from analysis any dropouts (i.e., patients permanently discontinuing their randomly assigned therapy before completing the 6-month course). In its advisory capacity, the DMC recommended that the investigators consider extending the planned 6-month treatment comparison period to get a better sense of the impact of peptide T, particularly its longer-term effect on cognitive impairment. It also advised an intent-to-treat type of analysis to minimize the risk of bias from excluding dropouts. The team of study investigators considered but did not accept the recommendation for longer-term follow-up. Their interaction with the patient community led them to believe that participants would not accept a longer period of time in which they might be receiving a placebo. The investigators did accept, however, the recommendation to analyze by intent-to-treat. Furthermore, after an extended discussion with the DMC, the NIMH program accepted the recommendation that the DMC would be responsible for ongoing review of both safety and efficacy data, revising its original concept that it would only review information on safety. Including efficacy data in the review would allow an evaluation of benefit-to-risk, of great importance when assessing appropriateness of trial continuation.

Example 2.3: The Cardiac Arrhythmia Suppression Trial

The Cardiac Arrhythmia Suppression Trial (CAST) was a placebo-controlled assessment of the safety and efficacy of anti-arrhythmia drugs (Cardiac Arrhythmia Suppression Trial, 1989; Echt *et al.*, 1991). The primary efficacy endpoint was patient survival. The trial was initially designed to use a one-sided 0.05 level of significance when testing for benefit. At its initial review of the protocol, the DMC questioned whether a one-sided 0.05 criterion was stringent enough for such an important pivotal study. The Steering Committee, accepting the advice of the DMC that such a trial should have a traditional 0.025 rather than 0.05 false positive error rate, agreed to base the test for benefit on the use of a one-sided 0.025 level of significance.

In addition to setting an 'upper boundary' to monitor the trial for conclusive evidence of treatment benefit, the DMC implemented a 'lower boundary' to monitor for conclusive evidence of treatment harm. This lower boundary was created to maintain an 0.025 probability of falsely concluding harm for a treatment truly having no effect on survival. In the CAST publication of primary trial results, this lower boundary was referred to as the advisory boundary for harm. As the

CAST trial evolved, this boundary for harm was crossed very quickly, despite the prior expectation for treatment benefit. The rapid detection of harm of these anti-arrhythmics was enhanced by the implementation of this lower boundary, in that the crossing of this boundary provided important reinforcement of the reliability of this negative result.

The examples in this section illustrate the potential value of DMC input into the design of the clinical trial. Although the role of the DMC is advisory and their recommendations are not always implemented by the study investigators and sponsor, the DMC can provide a fresh viewpoint and important insights about possible improvements to the protocol. If the protocol team (i.e., the trial's study investigators and sponsor) are unwilling to address ethical or scientific issues that the DMC members believe to be of critical importance, the members can consider their option to resign. Fortunately, such conflicts are rare. In settings in which the protocol team has relatively limited expertise in some important clinical or statistical areas of trial design or conduct, this advisory input by the DMC may be particularly helpful.

2.2.1.2 Review of procedures to ensure quality of study conduct

Capturing relevant information of high quality is a necessity if a clinical trial is to provide reliable and definitive conclusions about the efficacy and safety of therapeutic interventions. The DMC should have confidence that procedures are in place to ensure that high-quality information will be obtained. Thus, prior to the initiation of the trial, the DMC should make itself aware of the data management and quality control procedures that will be used.

The formation of the NIAID-sponsored AIDS Clinical Trials Group (ACTG) and its DMC provides an illustration of these issues (DeMets *et al.*, 1995). The complexities of clinical trials in AIDS, which have been greatly increased by intense pressure for obtaining rapid answers, made the development of appropriate systems and procedures within the ACTG for data management and quality control even more complicated than usual. Initially, the ACTG DMC had considerable difficulties understanding how data were being managed, what information on safety and efficacy could be provided on a timely basis, and how accurate it would be. After the DMC spent much of its first two quarterly meetings in early 1987 probing issues on data quality, NIAID decided to arrange for some DMC members to make a site visit of the AIDS Clinical Trials Data Coordinating Center (DCC).

This visit had a number of beneficial effects, for the DCC as well as for the DMC. First, it gave the latter a chance to convey directly to the former its sense of its own mission. The DCC had been inundated with extensive data requests from investigators, regulators, NIH programs and even members of Congress, and this meeting gave the DMC the opportunity to put into focus those data elements of particular importance to it in its review of ongoing studies. It provided the DMC a better understanding of DCC systems and procedures, and the opportunity to give some recommendations for improvements. Finally, it allowed the committee to work with the DCC to set up a unified format for the interim reports that the DMC would review.

It is not common for DMC members to visit the study's data coordinating center. It was thought to be necessary in the ACTG setting due to the commitment of this single committee to monitor a broad range of studies over a long period of time. It also was thought to be necessary due to the complexities of HIV/AIDS clinical trials that resulted from a newly emerging life-threatening infectious disease that was rapidly spreading, investigators who were relatively inexperienced with chronic disease clinical trials, and enormous external pressures. In contrast, after NIMH formed the DMC to monitor the single peptide T trial that was discussed in Example 2.2, that committee's review of the data systems and procedures was limited to careful discussions with investigators at each of its quarterly meetings. This was sufficient due to the less complex data management structures planned for that single trial.

It is difficult to provide specific rules regarding the extent to which a DMC needs to be involved in the sort of data management/report generation issues discussed above. Different settings will require different considerations. The key requirement is that the DMC be confident that adequate procedures are in place for data management and quality control. Without such confidence the committee cannot assume the interim data are reliable and will not be able to fulfill its mission of protecting study participants in its evaluation of these data.

2.2.2 Evaluating the quality of ongoing study conduct

A second opportunity for the DMC to assist in ensuring the quality of the clinical trial arises early during the conduct of the study. As with DMC recommendations regarding clinical trial design issues, their monitoring of quality of conduct issues can be very useful, particularly in settings in which the protocol team has relatively limited expertise in some important clinical or statistical areas of trial conduct.

The DMC will generally review the rate of accrual of study participants to see if it will allow timely completion of the trial. If the accrual rate is slower than anticipated, the DMC might simply express concern, or might suggest that the study team probe possible solutions. Such suggestions might be recruitment of additional study sites, improved promotion of the trial within existing sites, streamlining of unnecessarily burdensome study procedures, removing unnecessary restrictions on use of concomitant medications, and ensuring that eligibility criteria are sufficiently inclusive to allow enrollment of all patients who might benefit from the experimental therapy. The DMC should also assess compliance with eligibility restrictions, and should determine whether meaningful imbalances in important patient characteristics exist between intervention groups.

Other features relating to quality of trial conduct that should be monitored by the DMC early in the trial include patient adherence with assigned therapy, and physician adherence with the protocol (overall, as well as by intervention and by study site). Accuracy and completeness of data capture should also be assessed. Relevant features include currency of data on primary and secondary outcome measures and other important data elements, missing data, and rates of loss to follow-up, overall and by study site. If irregularities were identified sufficiently early in the trial, corrective actions usually would be possible.

While the study investigators, sponsor and steering committee will also be monitoring the quality of study conduct, the DMC will be able to perform a more comprehensive and integrated review due to its unique unblinded access to many of the relevant data elements. The DMC can also provide the benefits derived from independent oversight. The next example illustrates how early monitoring can be very important in providing timely identification of problems in trial conduct.

Example 2.4: The Nocturnal Oxygen Therapy Trial

In the NOTT trial (Nocturnal Oxygen Therapy Trial Group, 1980) the effect of continuous versus nocturnal oxygen supplementation on survival and other morbidity measures was tested in patients with advanced chronic obstructive pulmonary disease. As the trial progressed, a key subgroup appeared to show a nominally significant effect in favor of continuous oxygen supplementation, with no substantial trend apparent in the remaining participants. The DMC requested that the data collection process be examined for completeness. Results from that examination revealed that one or two clinical centers had been tardy in getting all patient data completed and had submitted more mortality data for the nocturnal oxygen treatment group than the continuous oxygen group. Although the trial was not blinded, there was no evidence that investigator bias caused the reporting delays; the problem seemed simply to be a random imbalance in the rate of data collection and reporting. When the data files were updated, the apparent difference in effect between subgroups had largely disappeared, having apparently been due to an artifact in data management (DeMets *et al.*, 1982).

A third opportunity for the DMC to have a positive influence on data quality arises during its conduct of formal interim data analyses. Through careful review of study reports, the DMC should attempt to determine whether proper standards for data completeness and accuracy have been implemented. One such standard will relate to the recency of follow-up available on all patients. It is desirable that the database be as current as possible, since interim analyses of the available data could lead to early termination of the trial. Some amount of time, however, must be allowed for editing and correcting the database, performing the interim analyses and transmitting the interim reports to the DMC members for review prior to the meeting. Experience from the perspective of both the statistical data center and the DMC member suggests that it is reasonable to expect the DMC to receive complete and accurate major endpoint information that is current for each study participant to within about 2 months of the date of the DMC meeting. This standard (i.e., allowing a lag of no more than 2 months for each study participant) would minimize the risk that the number of outcomes that would have occurred by the time of the DMC meeting but would not be available to the DMC would substantially alter conclusions from those data presented to the committee. As an example, in an HIV/AIDS study discussed in detail in Chapter 6, reducing the lag from 3 months to less than 1 month provided 50% more study endpoints and an altered perspective on the nature of the treatment effect. A detailed strategy to achieve currency of an interim database to within 2 months of the DMC meeting date is presented in Chapter 6.

The DMC should also critically assess the quality and accuracy of the reports they are provided to review at formal interim analyses. The importance of this aspect of review is illustrated by a recent trial.

Example 2.5: A trial with suspicious patterns at a formal interim analysis

A major international placebo-controlled trial of an intervention in an acute care setting had achieved approximately one-half its targeted endpoints and thereby had reached a protocol-specified milestone for a formal interim analysis. Detailed statistical guidelines for early termination were in place in the trial's DMC charter (see section 2.3), not only for settings in which interim results were strongly favorable, but also for when such results were strongly unfavorable.

At this formal interim analysis, the primary efficacy endpoint data were sufficiently negative that the confidence interval for treatment effect on the primary efficacy endpoint ruled out the level of treatment benefit that the trial was targeted to detect. The boundary for termination of the study as a negative trial formally had been crossed. According to trial guidelines, this result should trigger a termination recommendation from the DMC to the study's lead investigators and sponsor (i.e., the trial's steering committee), unless other evidence would provide strong rationale not to do so.

While there was a temptation to quickly arrive at a recommendation to terminate the trial, the DMC spent considerable time probing the interim data in the study reports. Some suspicious patterns were detected, including a few important laboratory and toxicity measures that revealed trends in the opposite direction to what might have been expected. The DMC had a dilemma. With such unfavorable results, it would be important to promptly inform the Steering Committee to enable them to spare future patients from being randomized to receive a potentially harmful treatment. On the other hand, with these patterns leading to the suspicion that the treatment code in the DMC reports might have been reversed, the DMC discussed delaying their recommendation to the steering committee until after the trial's data management group could examine study vials and treatment codes to ensure the accuracy of the DMC reports?

The DMC elected to delay. They met with the large group of international investigators and sponsor representatives who had gathered in a separate meeting room and announced to them that the DMC's recommendations about the trial would be delivered to them by teleconference a week later.

During the next week, after the trial's data management group had reviewed the content of study vials and the treatment codes, the DMC was informed that their suspicion that the treatment codes in the DMC report had been reversed had been confirmed. Corrected analyses showed results favoring the experimental treatment, with these interim data not crossing a boundary for termination. The DMC promptly held a teleconference to deliver their recommendation to the steering committee to continue the trial. The critical review of the data by the DMC averted the considerable disruption to the trial that would have occurred had a recommendation for termination for negative results been made.

2.2.3 Assessing safety and efficacy data

The most important responsibility of the DMC is the performance of ongoing reviews of the evolving safety and efficacy data by intervention group. Such reviews are particularly critical to safeguarding the interests of study participants, and should begin with early safety/trial integrity reviews, as discussed in Chapter 6. Very early assessments of safety parameters are especially important in settings where there would be considerable risk of rapidly emerging adverse events. It is widely recognized that efficacy and safety data also should be carefully reviewed at protocol-specified times of formal interim efficacy analyses.

At those meetings at which the study protocol and DMC charter have defined the main objective to be safety monitoring, the committee should give careful consideration to a comprehensive summary of evolving safety data. It also is important, although less uniformly recognized, that efficacy data should also be available for DMC review at such times. An intervention that introduces safety risks could still provide a favorable benefit-to-risk profile if it also provides important beneficial effects on efficacy. This is especially true in trials having major clinical primary endpoints, such mortality or serious morbidity. Of course, one can only address risk in the context of benefit by evaluating both of these components. Thus, even when one expects efficacy data to be too limited to potentially establish conclusive evidence of benefit, a review of efficacy data should be included at those interim analyses where safety data reviews could lead to recommendations for significant alterations or termination of the trial. Inclusion of limited efficacy data in such safety reviews, when using conservative group sequential monitoring procedures such as the O'Brien and Fleming (1979) guideline, will have a negligible effect on boundaries for significance that will be used later in the trial.

Following any interim review of efficacy and safety data, whether an early safety/trial integrity review or a formal interim efficacy analysis, the DMC should provide recommendations to the study investigators and sponsor about whether to continue or terminate the clinical trial, in addition to previously discussed recommendations about modifications in study design or in procedures for data management, quality control or reporting. A recommendation to terminate a trial would be made if the DMC judged the results to be convincingly positive or negative, or if it judged that the trial would be unable to conclusively answer the primary questions it was designed to address. Illustrations are provided for trials in which recommendations for termination were made due to favorable benefit-to-risk, unfavorable benefit-to-risk, or inability to answer the primary questions.

2.2.3.1 Termination due to favorable benefit-to-risk

A large majority of trials monitored by DMCs do not terminate early, but continue to their protocol-specified time of completion. However, interim efficacy results for some trials can be so compelling for clinically relevant outcome measures that early termination is recommended by an independent DMC. For example, three randomized placebo-controlled double-blind trials evaluating different beta-blocker drugs in chronic heart failure were terminated early due to highly statistically significant reductions in mortality and in mortality plus hospitalization (see Hjalmarson *et al.*, 2000; MERIT-HF Study Group, 1999; CIBIS-II Investigators and Committees, 1999; Packer *et al.*, 2001). Chronic heart failure is a progressive disease with an annual mortality rate of 15–20% in the more advanced stages. Beta-blockers had been avoided for over two decades in heart failure patients due to the effect of slowing the heart rate and lowering blood pressure, effects which were thought to make heart failure worse. However, some data did exist which suggested benefit. Thus, these trials were initiated. All three trials were industry-sponsored and had independent DMCs as well as independent statistical centers (see Chapter 7) to provide the DMC interim reports. Sponsors were blinded until the DMC recommended early termination. Each trial used group sequential boundaries (discussed in Chapter 8) to assess the strength of evidence for benefit before making their recommendation. In each case, interim comparisons of survival curves indicated nominal p-values of the order of 0.00001 (equivalent to a standardized statistic of 4.0).

A landmark study regarding the prevention of the spread of HIV further illustrates the importance of considering early termination due to compelling evidence for benefit.

Example 2.6: Prevention of vertical transmission in HIV-positive pregnant women

ACTG 076 was a placebo-controlled trial designed to determine whether antiviral therapy (AZT) would reduce the risk of transmission of HIV from an infected pregnant woman to her infant (Connor *et al.*, 1994). AZT or placebo was provided to the mother during the third trimester of pregnancy and intensively during labor/delivery, and to the infant for 6 weeks after delivery.

Recruitment began in April 1991, with a target enrollment of 748 mother–infant pairs. At the first formal interim efficacy analysis on February 17, 1994, the DMC was presented data on 421 infants born to the 477 women enrolled by December 20, 1993. Viral culture data was presented for the 364 infants from whom it had become possible to obtain a culture assessment. These data provided evidence of a striking threefold difference in HIV infection rates in the infants. In the placebo group, 40 of 184 infants (25.5%) had at least one positive culture while, in the AZT group, only 13 of 180 (8.3%) were culture positive. The log-rank test yielded $p = 0.0001$, with this strength of evidence clearly meeting the prespecified O'Brien–Fleming statistical guideline.

Because AZT was provided to infants until 6 weeks of age, the DMC required evidence that the drug was actually reducing the risk of HIV infection, rather than simply masking the infection in viral cultures obtained during the infant's first 24 weeks of life. The evidence that no additional cases of positive viral cultures were discovered in the 70 infants followed between 24 and 72 weeks was important in the DMC's deliberation about the convincingness of trial results.

At the time of this analysis, there had been seven infant deaths in each group, and it was too early to document the clinical benefit to the AZT infants in terms of reduction in occurrence/severity of AIDS-related illness and prolongation of survival. Similarly, no information was yet available about the potentially toxic effects of AZT on these infants, such as adverse effects on growth and development and immune function, as well as neurologic complications possibly resulting from AZT exposure *in utero*.

Due to the convincing nature of the effect of AZT in reducing the risk of vertical transmission of HIV, however, the DMC recommended that these interim results be released publicly, that study accrual be terminated, and that placebo group infants less than 6 weeks of age be offered AZT. Because considerable uncertainty about the totality of long-term risks and benefits of AZT for these infants still remained, the DMC urged that long-term follow-up continue for all mother–infant pairs. Such follow-up would enable documentation of the long-term effects of AZT on AIDS events and death, and long-term risks of neurological and other complications. This long-term follow-up proved to be particularly important in addressing subsequent concerns about the potential for increased risk of mitochondrial dysfunction associated with this regimen.

2.2.3.2 *Termination due to unfavorable benefit-to-risk*

Two trials are presented to illustrate the considerations for early termination due to lack of benefit.

Example 2.7: Trimetrexate in HIV-positive patients

The ACTG 029/031 trial, evaluating trimetrexate with leucovorin rescue (TMTX) as a treatment for pneumocystis pneumonia, was terminated due to unfavorable efficacy results after approximately two-thirds of the planned accrual had been achieved (Sattler *et al.*, 1994). The intention of the trial had been to establish that the efficacy of TMTX was at least equivalent, if not superior, to that achieved by the standard-of-care trimethroprim–sulfamethoxazole (T-SMZ) regimen. However, at the third of five planned analyses by the DMC, the estimated survival on TMTX was inferior to that on the T-SMZ control regimen, with the estimate of the TMTX / T-SMZ relative risk of death being 1.75.

Noting that the O'Brien–Fleming guideline for significance at this third of five planned interim analyses was 0.01, the DMC generated the nominal 99% confidence interval for the TMTX/T-SMZ relative risk, yielding (0.839, 3.65). These results conclusively ruled out the possibility that TMTX would provide meaningful survival improvement relative to T-SMZ. Specifically, the lower limit of the hazard ratio, 0.839, allows one to reject the hypotheses that the relative risk of death on TMTX is only 84% (or less) that on T-SMX. Secondly, even if the study were continued to its maximum accrual, it was determined that it would be extremely unlikely that final results would become sufficiently favorable even to rule out that survival with TMTX would be meaningfully inferior (i.e., to rule out TMTX has a death rate at least 25% higher) relative to that with T-SMZ, since that would require the upper limit, 3.65, to be reduced to approximately 1.25.

Finally, this decision for early termination was supported by practical considerations about difficulties with accrual, predicted by the chairs of the protocol team to grow worse, for a variety of reasons, during the upcoming months if the trial were to continue.

Example 2.8: Erythropoietin in hemodialysis patients with congestive heart failure

Cardiac disease patients undergoing hemodialysis were randomized between an experimental strategy of maintaining hematocrit levels at 42% using high doses of erythropoietin, against a control strategy of maintaining hematocrit levels at 30% using standard erythropoietin doses. The originally planned sample size of $n = 1000$ was increased to $n = 1500$ early in the trial, largely due to concerns about lower than expected rates of death or non-fatal myocardial infarction (MI), the trial's primary endpoint.

The DMC met in June 1996 to conduct its third interim analysis. Data were available through March 31, 1996, and provided approximately one-half of the trial's planned primary endpoints. At this analysis, unfavorable results were observed for the high-dose erythropoietin regimen (Besarab *et al.*, 1998). Specifically, the patients randomized to the experimental strategy involving higher doses of erythropoietin had a substantially *higher* death/MI rate – 202/618

(32.7%) vs. 164/615 (26.7%). The Kaplan–Meier estimates of the proportion of patients having either death or non-fatal MI at 18 months were 37% vs. 29% in the high-dose and standard-dose erythropoietin regimens, respectively. In addition to its adverse effect on the trial's primary endpoint, the high-dose erythropoietin regimen was associated with a higher death rate (195/618 vs. 160/615), and a significant increase in the frequency of thrombosis of the vascular access sites (243/618 vs. 176/615; nominal $p = 0.001$).

Statistical analyses using the method of repeated confidence intervals (Jennison and Turnbull, 1990) were performed to account for the three interim analyses that had been conducted using the Lan–DeMets implementation of the O'Brien–Fleming group sequential guideline (see Chapter 8). The estimated relative risk for death/MI, 1.30, indicated that the high-dose erythropoietin regimen provided a 30% increase in the rate of study endpoints, while the 95% repeated confidence interval, (0.94, 1.8), indicated one could rule out a greater than 6% reduction in the rate of death/MI attributable to the high-dose regimen. Since these results ruled out benefit and essentially came close to establishing harmful effects, early termination of the trial was recommended.

2.2.3.3 *Termination due to inability to answer trial questions*

Two trials are presented to illustrate issues that motivate early termination when there is inability to answer questions the trial was designed to address.

Example 2.9: Pyrimethamine for prevention of toxoplasmic encephalitis

A randomized, double-blind, placebo-controlled clinical trial was conducted to evaluate pyrimethamine in the prevention of *Toxoplasma gondii* encephalitis (TE), a serious brain infection, in HIV-infected individuals who were seropositive for this organism (Jacobson *et al.*, 1994). Survival was also a primary outcome. By March 1992, 396 patients had been randomized, 254 to the treatment arm and 132 to the placebo arm. At the time of review, 12 TE events had been observed – nine on pyrimethamine and three on placebo. The rate for the placebo arm (3.7/100 person years) was approximately one-third that expected, substantially reducing the power to detect a meaningful treatment benefit. Notably, many subjects in the trial were choosing to take ancillary treatment for prophylaxis of pneumocystis pneumonia. Although use of this treatment was not determined through randomization, it appeared to be associated with a considerable reduction in both TE and death rates. In addition, at the time of the DMC meeting, patients receiving pyrimethamine had higher death rates (34 deaths; 21.8/100 person years) compared to placebo (12 deaths; 14.4/100 person years). On March 17, 1992, the DMC recommended that the trial be terminated due to the unexpected low TE event rates that compromised the power of the trial, and due to early unfavorable trends in survival. This recommendation was accepted and implemented by

NIAID on March 30, 1992. Interestingly, subsequent updated analyses revealed the higher rate of death in the group receiving pyrimethamine not only persisted, but substantially increased (46 deaths, 27.8/100 person years versus 13 deaths, 14.5/100 person years).

Example 2.10: HIVIG and the prevention of vertical transmission of HIV

A randomized double-blind trial was conducted to evaluate the use of HIV-immune globulin (HIVIG) for prevention of maternal-to-child transmission of HIV in women and newborns receiving AZT (Stiehm *et al.*, 1999). The study target accrual was 800 mother–infant pairs, providing adequate power to detect a 50% reduction in the transmission rate under the assumption that the rate in the control regimen would be 15%.

At the interim analysis in January 1997, midway through the accrual period, the transmission rates were observed to be only 4.7% on both regimens. Not only were these just one-third the protocol-projected rates for the control arm, they were also declining over time. Specifically, among the 109 infants whose mothers were accrued in 1996 and who had viral cultures for HIV post delivery, only 2 (1.8%) were found to be HIV-positive.

With the low transmission rates and the lack of differences between intervention groups, detecting any advantage of HIVIG treatment would have been very difficult. Even if the sample size had been doubled to 1500, HIVIG would have to have had a true efficacy of 62% to yield 80% probability of achieving statistical significance in the overall final analysis. This fact, together with the declining rates of HIV transmission (possibly due to maternal exposure to increasingly effective concomitant antiretroviral regimens), led to a recommendation for study termination.

2.2.3.4 Continuation of ongoing clinical trials

We have reviewed many clinical trials that illustrate circumstances justifying a recommendation for early termination. Yet, it should be emphasized that a large majority of trials monitored by DMCs proceed, without early termination, to accrue the full information specified by the original design. Indeed, very often, the most important benefit of engaging a DMC, guided by proper statistical monitoring procedures, is to reduce the risk of inappropriate early termination based on prejudgment of unreliable early trends in efficacy and safety data.

The CPCRA 002 trial (Abrams *et al.*, 1994; Fleming *et al.*, 1995), discussed in Chapter 1, and the Coronary Drug Project trial (Coronary Drug Project Research Group, 1981), provide compelling illustrations of the important influence a DMC can have in avoiding inappropriate early termination of a trial. An equally compelling illustration, discussed in Chapter 4, is the Betaseron trial in patients with multiple sclerosis. This trial was continued to its prespecified time for completion, in spite of evidence for benefit that had been released from a concurrent related trial.

An example from the setting of diabetes management also is informative.

Example 2.11: The Diabetes Complications and Control Trial

The Diabetes Complications and Control Trial (DCCT) evaluated tight glucose control versus standard glucose control methods in an adult diabetic population (Diabetes Control and Complications Trial Research Group, 1993). The goal of the intervention was to reduce the occurrence of worsening diabetic retinopathy; the trial's primary outcome variable was the level of this worsening.

Among several secondary variables was presence of substantial microaneurisms. Since microaneurisms are predictive of proliferative retinopathy, some investigators had initially advocated using this variable as the primary endpoint to reduce trial's size and duration. However, in the final trial design, the DCCT investigators and DMC ultimately designated worsening retinopathy as the primary endpoint, recognizing that microaneurism had not been established to be a reliable surrogate for the occurrence of worsening proliferative retinopathy.

Early in follow-up, the DMC observed adverse trends in the presence of microaneurisms, trends favoring the standard glucose control regimen. However, the DMC waited to determine whether the primary outcome variable tracked the microaneurism endpoint. After further follow-up, the trend in microaneurisms reversed direction, with the tight glucose control regimen then having a decrease in microaneurisms. Ultimately, the trial provided significant evidence that the experimental regimen of tight glucose control did lead to a reduction in occurrence of worsening in proliferative retinopathy.

The DMC's recommendation to continue the DCCT trial, in spite of early unfavorable evidence relating to the microaneurism surrogate endpoint, enabled the trial to provide much more reliable and favorable assessments regarding the true efficacy of the tight glucose control regimen. With these long-term insights about favorable effects on occurrence of worsening in proliferative retinopathy, the DMC eventually did recommend that the protocol team terminate the trial prior to its prespecified time for completion.

The following example provides an illustration of a trial that was continued to its prespecified time for completion, in spite of important interim evidence of an unfavorable effect on mortality for the experimental intervention.

Example 2.12: The Heart and Estrogen/Progestin Replacement Study

The Heart and Estrogen/Progestin Replacement Study (HERS) trial evaluated the potential cardiovascular benefits of hormone replacement therapy (HRT) in postmenopausal women with coronary disease (Hulley *et al.*, 1998). HERS was a randomized double blind placebo-controlled trial sponsored by industry. The trial had an independent steering committee, an independent DMC, and an independent statistical center (see Chapter 7). The sponsor was blinded during the conduct of the trial.

HRT is a widely used therapy for symptom relief in postmenopausal women. Observational data suggested a cardiovascular benefit of HRT, but HERS was the first randomized trial to address survival benefit. As the trial progressed, a non-significant but negative (i.e., harmful) trend in survival began to emerge. Although the negative trend was worrisome to the DMC, it was important to distinguish between a neutral result and a true negative effect; a neutral result on mortality would still be consistent with HRT use for symptom relief, whereas a negative result in mortality would likely provide support for not using HRT for this purpose. As HERS progressed, the negative trend did not continue and when the trial was completed, the mortality results were nearly identical after a year of follow-up. Nevertheless, an early adverse effect on thrombotic events was still apparent. The continuation of HERS permitted distinguishing between an overall adverse mortality effect and what appears with further follow-up to be a neutral effect.

2.2.3.5 Consideration of the overall picture: primary and secondary analyses

In well-designed phase III clinical trials, the primary endpoint should be that outcome having the greatest clinical relevance to the patient that also will be particularly sensitive to the anticipated effects of the intervention. For example, while improving the duration of survival may often be of greatest relevance in the setting of life-threatening diseases, one might select another primary endpoint (such as pain relief for an analgesic treatment) when mortality is not expected to be favorably or adversely affected. Typically, then, the entire false-positive error rate is 'spent' on that primary endpoint, with group sequential procedures designed to allow interim monitoring while preserving the desired error rate.

While recommendations about trial termination or continuation should be based in the main on the primary efficacy endpoints and corresponding group sequential guidelines, the formulation of these recommendations must also include a global consideration of all available data, including secondary endpoints, toxicities, and data quality, as well as relevant information external to the trial. Such considerations could lead to continuation of trials when efficacy boundaries are crossed or termination when such boundaries are not crossed, particularly when strong results are obtained relating to secondary endpoints of considerable clinical relevance. The ACTG 081 Trial provides an important illustration of this issue.

Example 2.13: Prevention of serious fungal infection

The ACTG 081 Trial accrued 424 patients between September 1989 and April 1992, to determine whether fluconazole would be more effective than clotrimazole troches in the prevention of serious fungal infection in AIDS patients (Powderly *et al.*, 1995). This accrual was expected to produce approximately 25 fungal infection events, adequate to provide high power to detect a reduction in the 18-month rates from 10% to 2.5%. At an interim analysis in early May 1992, patients receiving fluconazole were seen to have experienced a striking reduction in the primary endpoint of serious fungal infections (2 vs. 14; $p = 0.0028$). Even though the O'Brien–Fleming boundary had clearly been crossed, the DMC recommended continuation of the trial after observing a large excess of deaths (45 vs. 31) on the fluconazole regimen. At the next meeting of the DMC in November 1992, the DMC again recommended continuation of follow-up for six additional months, since there was persistence of beneficial effects observed relative to serious fungal infections (4 vs. 18; $p = 0.0022$) and of adverse effects on mortality (66 vs. 51; $p = 0.088$) possibly arising from unintended interactions with other prophylaxis regimens.

At the completion of the follow-up period, on June 30, 1993, the fluconazole patients continued to show a lower rate of serious fungal infections (9 vs. 23; relative risk = 0.3; $p = 0.02$) and a higher rate of deaths (98 vs. 89; relative risk = 1.1), although the strengths of these associations were reduced from what had been observed at the interim analysis). Interestingly, the number of patients who experienced either a serious fungal infection or death continued to be higher on the fluconazole arm (102 vs. 96). The greater number of fluconazole patients who died without experiencing a serious fungal infection after randomization, (93 vs. 73) did suggest the need for further exploration of a possible unintended mechanism of fluconazole that could adversely affect survival.

2.2.3.6 Modifying sample sizes based on ongoing assessment of event rates

In clinical trials designed to detect a reduction in the relative risk (in time-to-event data) or in the odds ratio (in dichotomous endpoint data), specification of the relative risk (or odds ratio), r, to be detected with a given level of power and with a specified false positive error rate, leads to a specification of the number of participants, L, who must experience the primary endpoint in the trial (see Fleming and Harrington, 1991, Exercise 4.7). In turn, by estimating the event rate in the trial, one can derive the sample size.

This event rate usually is not known with precision at the time of trial planning. It is often necessary to revise sample size calculations in the early to middle stages of the trial when information on the event rate allows a more accurate calculation of the sample size that would be necessary to achieve L events in the trial. When sample size revisions are necessary, most often it is because the event rate has been overestimated in the protocol, leading to the need to increase the sample size from the original calculations in order to restore the power intended in the original design.

There are several different approaches to revising sample sizes while a trial is ongoing, and it is beyond the scope of this chapter to address them in any detail. Traditional approaches base such revisions on the overall event rate, or the variance of the outcome measure (Wittes and Brittain, 1990); newer approaches have proposed basing sample size changes on an interim estimate of the treatment effect (Lehmacher and Wassmer, 1999; Cui *et al.*, 1999; Wang *et al.*, 2001). Little experience is yet available with these new approaches; it is too early to assess whether and in what circumstances they will be useful. In any case, it is important that any sample size revisions made at an interim point in the trial be based on a prespecified, statistically justifiable plan that avoids bias. Otherwise, it will be difficult to interpret the statistical comparisons that will eventually be performed. It is therefore important for the sponsor and/or steering committee to develop a clear algorithm for when and how these revisions will be carried out, and describe this in the protocol.

A clinical trial in children with meningococcemia provides an illustration.

Example 2.14: BPI in infants with meningococcemia

Severe meningococcal disease imparts significant morbidity and mortality risks in pediatric populations. A randomized trial of bactericidal/permeability-increasing protein (BPI) was conducted in 1996–1999 to determine whether BPI would improve survival in children with this disease (Levin *et al.*, 2000). Since the mortality risk is acute in this setting, the primary endpoint of the trial was 60-day mortality.

Pilot data for the planned phase III trial were provided by a single-arm phase II trial together with a matched historical control group. In the phase II trial, the 60-day mortality rate for 22 children receiving BPI was 5% (1/22). In the matched historical control group, the 60-day mortality rate was 24% (10/42). Based on these pilot data, the phase III trial was designed to provide 85% power to detect an odds ratio of 3 (25% vs. 10%) for the 60-day mortality rate on placebo relative to BPI. To achieve this power, the trial would need to accrue and follow children until $L = 35$ deaths occurred.

With the assumed 25% and 10% death rates on placebo and BPI, it was estimated that a sample size of 100 per group would be required. However, at the third meeting of the DMC in March 1998, the death rate in the pooled sample was only 11.5% (18/157). Hence, the DMC recommended an increase in sample size to 150 per arm. Interestingly, the death rate in children enrolled into the trial continued to noticeably decrease. Between the DMC's fourth meeting in September 1998 and fifth meeting in March 1999, when enrollment approached a total of 300 children, the death rate in the pooled sample approached only 10%, leading to a recommendation to increase sample size to 175 per arm. At its final meeting in March 1999, with 33 deaths in 339 enrolled children, the DMC recommended enrollment continue until June 1999 unless 35 deaths occurred at an earlier date.

These recommendations relating to increases in sample size occurred at rather late stages of this trial due to a surprising decreasing trend in the death rate of children who enrolled. However, integrity of the monitoring process was preserved since it had been clearly specified that the defining and constant element of the accrual goal was to achieve $L = 35$ deaths rather than a given sample size computed before trial initiation. With $L = 35$ deaths, the trial would provide 85% power to detect an odds ratio of 3 whether the true death rates were 25% vs. 10% (corresponding to the initial projection of a pooled event rate of 17.5%), or 17% vs. 6.4% (corresponding to the pooled event rate of 11.7% as seen at the third DMC meeting), or 14.6% vs. 5.4% (corresponding to the pooled event rate of 10% seen before the fifth DMC meeting.)

In this example, the DMC monitored the interim event rates and recommended sample size increases based on the prespecified algorithm. Since the DMC has access to the event rates it might appear that the DMC would be the logical entity to implement the algorithm and make such recommendations. In fact, it has been reasonably common practice for DMCs to assume this responsibility. On the other hand, because the DMC has access to the comparative data, one could also argue that the DMC should not be involved in making judgments about sample size. A DMC observing, for example, a lower than expected event rate but no trend suggesting that the null hypothesis will ultimately be rejected may be reluctant to recommend that the study be enlarged, even when the possibility of an ultimate difference being shown cannot be ruled out. Similarly, a DMC observing a lower than expected event rate but a larger than expected treatment effect, such that early termination at a future review might be in the offing, might also be reluctant to recommend enlarging the study.

Alternatively, the statistician preparing the interim reports could be charged with notifying the trial leadership if the overall event rate (or the event rate in the control group, whichever is the preferred basis for evaluating sample size in that trial) is falling substantially below the expected level, with the specific threshold to be established prior to any interim review. Such an approach could trigger implementation of the prespecified algorithm for revision of sample size without unblinding trial leadership to the comparative rates.

2.2.4 Reviewing the final results

After the completion of the follow-up and any final changes to the study database, the protocol team may proceed with its review of unblinded data. Because of the unique insights obtained by the DMC through its unblinded review of efficacy and safety data throughout the conduct of the trial, members of the DMC may in some cases be invited to discuss their interpretation of final results with the study investigators and sponsor. This discussion could be especially informative to the investigators and sponsor in settings in which the DMC had recommended substantial changes to the trial conduct during the course of the study. This review of final results will be maximally informative if the investigators and sponsor are provided access to the minutes of the DMC deliberations to more fully understand the insights the committee had during its monitoring of the trial.

While additional involvement of the DMC at the end of the trial could be valuable to the protocol team, it generally is not the role of the DMC membership to participate in detailed data analyses after trial completion. Although identification of the membership and role of the DMC should be provided in manuscripts, in order to maintain the DMC's independence it generally is inappropriate for members to be authors of the manuscript that provides the primary results of the trial.

2.3 THE DATA MONITORING COMMITTEE CHARTER

The DMC Charter should provide a detailed presentation of the membership, role and responsibilities of the DMC and, if relevant, of steering committee members. It should indicate the timing and purpose of DMC meetings, the procedures for maintaining confidentiality, and the format and content of DMC reports (see Chapter 6.) The Charter should specify the statistical procedures, including the monitoring guidelines, which will be used to monitor the identified primary, secondary and safety outcome variables. For example, if a group sequential procedure is to be used, then tables or figures of the monitoring boundaries might be provided. Plans for changing frequency of interim analyses may be described as well as procedures for recommending protocol changes, such as increasing the sample size or duration of follow-up when event rates are unexpectedly low. Any subgroup analyses of key interest should also be indicated. A sample charter can be found in Appendix A.

Statistical methods for interim monitoring serve primarily as guidelines rather than rules, because such methods cannot capture all of the issues that a DMC must consider (Coronary Drug Project Research Group, 1981). Thus, in addition to the statistical details, the DMC Charter should also outline some of the factors, beyond the strength of evidence about treatment effect on the primary outcome, that should be weighed in developing recommendations about trial continuation. Some of these are listed in Table 1.1.

For example, early in a trial when sample sizes are small, the baseline risk factors between treatment arms may not yet be well balanced between treatment groups. Such imbalances must be considered in any evaluation of either effectiveness or safety. Incompleteness of data, especially for important outcome measures, can also introduce biases that could lead to misleading trends. Trends that are not internally consistent, such as effects emerging in some patient subsets or study sites but not others, are more difficult to evaluate and often cause DMCs to delay decisions to see whether these inconsistencies are eventually resolved.

REFERENCES

Abrams D, Goldman A, Launer C *et al.* (1994) A comparative trial of didanosine or zalcitabine after treatment with zidovudine in patients with human immunodeficiency virus infection. *New England Journal of Medicine* **330**: 657–662.

Besarab A, Bolton WK, Browne JK *et al.* (1998) The effects of normal as compared with low hematocrit values in patients with cardiac disease who are receiving hemodialysis and epoetin. *New England Journal of Medicine* **339**: 584–590.

Cardiac Arrhythmia Suppression Trial (1989) Preliminary report: Effect of encainide and flecainide on mortality in a randomized trial of arrhythmia suppression after myocardial infarction. *New England Journal of Medicine* **312**: 406–412.

CIBIS-II Investigators and Committees (1999) The cardiac insufficiency bisoprolol study II (CIBIS-II): A randomised trial. *Lancet* **353**: 9–13.

Connor E, Sperling R, Gelber R *et al.* (1994) Reduction of maternal-infant transmission of human immunodeficiency virus type-1 with zidovudine treatment. *New England Journal of Medicine* **331**: 1173–1180.

Coronary Drug Project Research Group (1981) Practical aspects of decision making in clinical trials: the Coronary Drug Project as a case study. *Controlled Clinical Trials* **1**: 363–376.

Cui L, Hung HMJ, Wang SJ (1999) Modification of sample size in group sequential trials. *Biometrics* **55**: 853–857.

DeMets DL, Williams GW, Brown Jr. BW and the NOTT Research Group (1982) A case report of data monitoring experience: the Nocturnal Oxygen Therapy Trial. *Controlled Clinical Trials* **3**: 113–124.

DeMets DL, Fleming TR, Whitley RJ *et al.* (1995) The data and safety monitoring board and acquired immune deficiency syndrome (AIDS) clinical trials. *Controlled Clinical Trials* **16**: 408–421.

Diabetes Control and Complications Trial Research Group (1993) The effect of intensive treatment of diabetes on the development and progression of long-term complications in insulin-dependent diabetes mellitus. *New England Journal of Medicine* **329**(14): 977–986.

Echt DS, Liebson PR, Mitchell LB *et al.* (1991) Mortality and morbidity in patients receiving encainide, flecainide, or placebo. The Cardiac Arrhythmia Suppression Trial. *New England Journal of Medicine* **324**: 781–788.

Fleming TR, DeMets DL (1993) Monitoring of clinical trials: Issues and recommendations. *Controlled Clinical Trials* **14**: 183–197.

Fleming TR, Harrington DP (1991) *Counting processes and survival analysis.* John Wiley & Sons, New York.

Fleming TR, Neaton JD, Goldman A, DeMets DL, Launer C, Korvick J, Abrams D, and the Terry Beirn Community Programs for Clinical Research in AIDS (1995) Insights from monitoring the CPCRA didanosine/zalcitabine trial. *Journal of Acquired Immune Deficiency Syndromes and Human Retrovirology* **10**(Suppl. 2): S9–S18.

Green SJ, Fleming TR, O'Fallon JR (1987) Policies for study monitoring and interim reporting of results. *Journal of Clinical Oncology* **5**: 1477–1484.

Heart Special Project Committee (1988) Organization, review and administration of cooperative studies (Greenberg Report): a report for the Heart Special Project Committee to the National Advisory Council, May 1967. *Controlled Clinical Trials* **9**:137–148.

Heseltine PN, Goodkin K, Atkinson JH, Vitiello B, Rochon J, Heaton RK, Eaton EM, Wilkie FL, Sobel E, Brown SJ, Feaster D, Schneider L, Goldschmidts WL, Stover ES (1998) Randomized double-blind placebo-controlled trial of peptide T for HIV-associated cognitive impairment. *Archives of Neurology* **55**(1): 41–51.

Hjalmarson A, Goldstein S, Fagerberg B, Wedel H, Waagstein F, Kjekshus J, Wikstrand J *et al.* (2000) Effects of controlled-release metoprolol on total mortality, hospitalizations, and well-being in patients with heart failure. The metoprolol CR/XL randomized intervention trial in congestive heart failure (MERIT-HF). *Journal of the American Medical Association* **283**(10): 1295–1302.

Hulley S, Grady D, Bush T, Furberg C, Herrington D, Riggs B, Vittinghoff E, for the Heart and Estrogen/Progestin Replacement Study (HERS) Research Group (1998) Randomized trial of estrogen plus progestin for secondary prevention of coronary heart disease in postmenopausal women. *Journal of the American Medical Association* **280**: 605–613.

International Chronic Granulomatous Disease Study Group (1991) A controlled trial of interferon gamma to prevent infection in chronic granulomatous disease. *New England Journal of Medicine* **324**(8): 509–516.

Jacobson MA, Besch CL, Child C *et al.* and the Community Programs for Clinical Research in AIDS (1994) Prophylaxis with pyrimethamine for toxoplasmic encephalitis in patients with advanced HIV disease: results of a randomized trial. *Journal of Infectious Diseases* **169**: 384–394.

Jennison CJ, Turnbull BW (1990) Statistical approaches to interim monitoring of medical trials: a review and commentary. *Statistical Science* **5**: 299–317.

Lehmacher W, Wassmer G (1999) Adaptive sample size calculations in group sequential trials. *Biometrics* **55**: 1286–1290.

Levin M, Quint PA, Goldstein B *et al.* (2000) Recombinant bactericidal/permeability-increasing protein (rBPI21) as adjunctive treatment for children with severe meningococcal sepsis: a randomised trial. *Lancet* **356**: 961–967.

MERIT-HF Study Group (1999) Effect of metoprolol CR/XL in chronic heart failure: Metoprolol CR/XL randomised intervention trial in congestive heart failure. *Lancet* **353**: 2001–2007.

Nocturnal Oxygen Therapy Trial Group (1980) Continuous or nocturnal oxygen therapy in hypoxemic chronic obstructive lung disease. A clinical trial. *Annals of Internal Medicine* **93**(3): 91–98.

O'Brien PC, Fleming TR (1979) A multiple testing procedure for clinical trials. *Biometrics* **35**: 549–556.

Packer M, Coats AJS, Fowler MB, Katus HA, Krum HA, Mohacsi P, Rouleau JL, Tendera M, Castaigne A, Staiger C, Curtin EL, Roecker EB, Schultz MK and DeMets DL for the Carvedilol Prospective Randomized Cumulative Survival (COPERNICUS) Study Group (2001) Effect of Carvedilol on survival in severe chronic heart failure. *New England Journal of Medicine* **334**(22): 1651–1658.

Powderly WG, Finkelstein DM, Feinberg J, Frame P, He W, van der Horst C, Koletar SL, Eyster ME, Carey J, Waskin H, Hooton TM, Hyslop N, Spector SA, Bozzette SA (1995) A randomized trial comparing fluconazole with clotrimazole troches for the prevention of fungal infections in patients with advanced human immunodeficiency virus infection. *New England Journal of Medicine* **332**: 700–705.

Sattler FR, Frame P, Davis R, Nichols L, Shelton B, Akil B, Baughman R, Hughlett C, Weiss W, Boylen CT *et al.* (1994) Trimetrexate with leucovorin versus trimethoprim-sulfamethoxazole for moderate to severe episodes of *Pneumocystis carinii* pneumonia in patients with AIDS: a prospective, controlled multicenter investigation of the AIDS Clinical Trials Group Protocol 029-031. *Journal of Infectious Disease* **179**(1): 165–172.

Stiehm ER, Lambert JS, Mofenson LM, Bethel J, Whitehouse J, Nugent R, Moye J Jr, Fowler MG, Mathieson BJ, Reichelderfer P, Nemo GJ, Korelitz J, Meyer WA, Sapan CV, Jiminez E, Gandia J, Scott G, O'Sullivan MJ, Kovacs A, Stek A, Shearer WT, Hammill H (1999) Efficacy of zidovudine and human immunodeficiency virus (HIV) hyperimmune immunoglobulin for reducing perinatal HIV transmission from HIV-infected women with advanced disease: results of Pediatric AIDS Clinical Trials Group protocol 185. *Journal of Infectious Disease* **179**(3): 567–575.

Wang SJ, Hung HMJ, Tsong Y, Cui L (2001) Group sequential strategies for superiority and non-inferiority hypotheses in active controlled clinical trials. *Statistics in Medicine* **20**: 1903–1912.

Wittes J, Brittain E (1990). The role of internal pilot studies in increasing the efficiency of clinical trials. *Statistics in Medicine* **9**: 65–72.

3

Composition of a Data Monitoring Committee

Key Points

- DMCs should be multidisciplinary, and should always include individuals with relevant clinical and statistical expertise.
- Different trials may require the inclusion of different disciplines on a DMC.
- The appropriate size of a DMC depends on the type of trial and types of expertise needed.
- The study sponsor either appoints the DMC, or delegates this responsibility to another group such as a steering committee.
- Training programs for DMC service are needed.

3.1 INTRODUCTION

Even the simplest clinical trial is a fundamentally complex enterprise. Those making important judgments as clinical trials progress – whether to recommend stopping or continuation, changes in procedures or design, notification of other parties or continuing to maintain confidentiality of interim data – must bring to bear on the action being considered the current safety and efficacy data, data on other relevant variables (such as laboratory measurements), the possibility of trend reversals, the implication of results of other trials, the quality and reliability of the database, the potential impact of the action on other ongoing trials, potential regulatory implications and many other issues. The process of making such judgments is rarely straightforward; the observation of a low *p*-value or a sequential boundary crossed represents only the beginning of this process, not the end.

In order for a data monitoring committee to adequately perform its functions, it must be able to call on all the different types of expertise necessary to design and carry out the trial. The necessary expertise will vary to some extent from trial to trial, and some special cases will be considered in Chapter 9, but there is much commonality.

3.2 REQUIRED AREAS OF EXPERTISE

Clinical expertise in the medical area being studied is the most obvious type of expertise needed by a DMC. Studies of treatments for myocardial infarction will require cardiologists on their DMCs, studies of stroke treatments will require neurologists, studies in glaucoma will require ophthalmologists, and so on. Specific expertise in the clinical field is necessary in order to interpret adverse outcomes (part of the disease process or an adverse reaction to treatment?), benefit-to-risk issues (should we continue the study if there is a positive trend on the primary endpoint but simultaneously emerging safety concerns?), implications of external data that become available during the course of the trial (how relevant are those results to our study?), and many other issues that may arise during the course of the trial. These considerations should include non-medical clinical expertise when appropriate; for example, psychologists may be included on a DMC for a trial in which psychosocial endpoints are considered (Wittes, 1993).

Statisticians are also an essential component of DMCs. There has been extensive development of statistical methods for monitoring clinical trials over the past 30 years (see Chapter 8), and there are a variety of different approaches that are considered generally acceptable. Most DMCs use such methods, which account for the multiple opportunities to assess the data and draw conclusions, to evaluate the strength of the accumulating data. The presentation of an interim analysis to a DMC by the trial statistician is often statistically complex, and may use new methods that are unfamiliar to committee clinicians, even those with substantial trials experience; statisticians are needed on the DMC to fully interpret these presentations and ensure that the analyses performed are adequate to support decision-making. Clearly, statisticians serving on such committees should be knowledgeable about statistical issues in clinical trials. This knowledge should be more than theoretical; personal experience as a clinical trials statistician who has been intimately involved with designing and implementing trials – developing randomization plans, quality control mechanisms, sequential monitoring approaches, analytical plans – is essential background for serving on a DMC. Statisticians with applied experience in the medical area under study can make particularly valuable contributions.

While statisticians frequently serve on other types of biomedical advisory panels, such as NIH Consensus Conference panels and FDA advisory committees, their role on DMCs is perhaps more central to the committee's function than it is for these other types of panels. The proportion of members who provide statistical expertise may be higher for a DMC than for other advisory groups; frequently, more than one statistician serves on a DMC, particularly for government-sponsored trials. The presence of multiple statisticians reduces the chance that important perspectives on the statistical interpretation of the data being reviewed are being overlooked, and promotes a fuller and more quantitatively informed discussion of the strength of the evidence at hand.

Virtually all DMCs include both clinicians and statisticians. In government-sponsored trials, bioethicists with knowledge of and experience with clinical trials are also frequently invited to serve on DMCs. The generally perceived role of the bioethicist is to serve as the patient advocate. While all the members of the DMC bear the charge of ensuring the continuing safety of patients entered onto the trial and treated according to the trial protocol, the bioethicist has as his/her fundamental responsibility to ensure that the scientific goals of the study, however meritorious, do not lead to actions that are unacceptable from the perspective of the study patients. The bioethicist may identify aspects of the study design that should be reconsidered, or that may need to be clarified in the informed consent document. In addition, as the trial progresses, the bioethicist may recommend changes to the informed consent, either on the basis of new information external to the trial or of emerging data from the trial being monitored.

The input of a bioethicist can be particularly valuable when unanticipated decision points arise. For example, an interim analysis may show that a new treatment aimed at relieving major symptoms of a chronic disease does seem to have the intended beneficial effect, yet there is an unexplained worsening of survival in the patients receiving this treatment. There is no 'right answer' as to how much worse the survival needs to be, or how large the beneficial effects need to be, with regard to making the decision about stopping or continuing the study. The perspective of the bioethicist, in our experience, can add substantially to the critical input of the clinicians and statisticians in such circumstances.

In some instances, representatives of the patient community are asked to serve on DMCs. The current guidelines of the National Cancer Institute (uniquely among NIH institutes) call for inclusion of a patient representative on DMCs for all cooperative cancer groups (Smith *et al.*, 1997). The purpose of including a patient representative is to bring into the discussion someone with direct experience as a patient, or as a close relative of a patient. Such individuals may bring a unique perspective to DMC discussions, and their inclusion acknowledges the fundamental partnership of patients and scientists in the research enterprise. This special perspective, however, brings with it potential concerns. First, while such individuals need not have extensive scientific backgrounds, they do need to have sufficient grounding in the fundamental principles and methods of clinical trials to be able to interpret the data they will be considering. Second, it may not be obvious how to go about identifying and selecting a 'patient representative', particularly in settings in which multiple advocacy groups have arisen. But the third concern is perhaps the most important: will it be too great a burden for such individuals to maintain confidentiality in cases where emerging results are impressive but not yet sufficiently definitive to warrant final conclusions? An individual whose role in an advocacy group involves providing advice to persons with the disease may find it problematic to ignore information available to him/her only as a DMC member. The problem will be even more intense if the individual in question had favored early stopping and release of data on the basis of an interim analysis but the committee as a whole ultimately recommended continuation.

Physicians on DMCs face an analogous problem, in that they are also in regular contact with individuals with the disease under study and must recommend treatment approaches for them, but the situations are substantially different. The role of a physician is to recommend appropriate treatment for patients, so it is not a surprise that a physician serving on a DMC recommends treatments to his/her patients. It would be more notable, and more suggestive of some special information, if a patient advocate known to be serving on a DMC suddenly began to recommend a particular treatment. It is possible that such concerns will prove to be groundless in all or most disease areas, and if this is the case it will become clear as we gain more experience with DMCs that include patient representatives.

(Physicians on DMCs are not entirely free of such problems, however. A DMC physician managing a hospital unit, for example, might come to believe, on the basis of interim data from the trial being monitored, that the standard practice in this unit should change. Implementing such a change, unlike recommending treatments for individual patients, would be visible to professional colleagues and would suggest something about the interim results of the trial.)

Other areas of expertise may be useful on certain committees. In a survey of approaches to monitoring trials sponsored by the NIH (Geller and Stylianou 1993), the inclusion of epidemiologists, lawyers, and clinicians whose expertise was considered to be 'clinical trials' rather than the particular medical area being studied was reported as 'usual' by one or more institute respondents. Some trials may require the expertise of a pharmacologist or toxicologist, particularly when less than the usual amount of preliminary clinical data has been collected prior to the initiation of the monitored trial. In some cases, basic science expertise – for example, in microbiology or biochemistry – may be valuable. One of the earliest trials to report data monitoring considerations, the Coronary Drug Project, included pharmacologists/toxicologists, a lipid chemist and a clinical chemist on its DMC (Coronary Drug Project Research Group, 1981). In trials of medical devices, engineering expertise may be important. The need for 'atypical' expertise on a DMC should always be considered.

In 1998, the NIH issued a policy for data and safety monitoring of clinical trials (National Institutes of Health, 1998). With regard to DMC membership, the document states:

> Monitoring activities should be conducted by experts in all scientific disciplines needed to interpret the data and ensure patient safety. Clinical trial experts, biostatisticians, bioethicists, and clinicians knowledgeable about the disease and treatment under study should be part of the monitoring group or be available if warranted.

An optimal DMC is one in which each member knows enough about each aspect of the trial – the clinical issues, the statistical issues, the ethical issues – that they can all contribute to all aspects of the discussion. If one were to observe a DMC consisting of good scientists with appropriate expertise who are working together effectively, one might have difficulty initially in guessing who represented which discipline. While it is not always possible to achieve this for every DMC, it is a reasonable goal to keep in mind in constructing such committees.

3.3 OTHER RELEVANT CHARACTERISTICS OF COMMITTEE MEMBERS

In addition to the types of expertise discussed above, there are other characteristics desirable in DMC members. Prior experience on a DMC is very valuable. While one would quickly run out of candidates for DMCs if every member of every DMC had to have prior experience, it is important to make sure that at least some DMC members have had prior experience. All DMC members should optimally have some knowledge and understanding of, and practical experience with, clinical trials. Committee members must be willing to commit to attendance at meetings and to the preparatory review of materials necessary to the conduct of a productive meeting; a DMC is no place for an 'honorific' appointment. Finally, personality considerations are also important. DMC members should be reasonably assertive types – they should not hesitate to share what they are thinking, or to raise questions when they do not understand something or have identified concerns. On the other hand, an individual who tends to 'take over' and who has a hard time engaging in a productive dialogue with someone holding an opposing view would not be an ideal DMC member.

DMCs for international trials should be international as well. For large trials in which patients are entered from many different countries, it may not be feasible to include someone from each participating country on the DMC, but the trial sponsor should make every effort to include a diversity of perspectives that will adequately reflect the trial population.

Because DMCs do not operate in the public eye, they have been less subject to consideration of demographic diversity than other advisory committees convened by health agencies. Such considerations may be important in some cases, however, particularly when the disease being studied and/or the community most at risk of that disease may be associated with a particular demographic subgroup. As an example, an all-male DMC making a controversial recommendation concerning a high-profile breast cancer study might be more heavily criticized than a gender-heterogeneous DMC making the same recommendation; such criticism could ultimately affect the credibility of the study.

3.4 COMMITTEE SIZE

The sizes of DMCs vary widely. Some of this variability relates to the specific circumstances of the trial, as considered below, but some is probably attributable simply to the parallel evolution of DMC practices in different settings. Wittes (1993) reported experience serving on monitoring committees of as few as 3 to more than 15 members. Hawkins (1991), discussing committees for trials sponsored by the National Eye Institute, reported that the median size was 10, with a range of 7 to 15. (These numbers include *ex officio* as well as appointed members.) Parmar and Machin (1993) report that DMCs for cancer studies conducted by the British Medical Research Council consist of only three members, two clinicians and a statistician. Policies for DMCs of the cooperative cancer groups sponsored by the National Cancer Institute specify that the size of these committees should be limited, and should generally include no more than 10 members (Smith, 1997). DMCs monitoring a single trial sponsored by a pharmaceutical company frequently include only three or four members. Overall, a DMC should be large enough for a diversity of views and representation of all needed expertise, but small enough to promote full engagement and participation of all members as well as facilitate the scheduling of meetings. Our own experience as participants on these committees suggests that a DMC should be as small (with a minimum size of 3) as can provide the necessary expertise for competent monitoring of the study.

Trials that are particularly complex, such as those that address multiple questions, involve a variety of treatment modalities or include a highly diverse population with special considerations required for different subgroups, may require a somewhat larger DMC to ensure adequacy of monitoring. For example, a DMC for a cancer treatment trial in which an immune system enhancer (such as a vaccine) is added to a standard chemotherapy–radiotherapy regimen may require an immunologist as well as one or more medical and radiation oncologists. A factorial trial in which two or more interventions are tested in the same population may also require more than the usual numbers of areas of expertise to be represented on the DMC.

DMCs that monitor multiple trials often need to be larger than those monitoring a single trial, for several reasons. Perhaps the most obvious reason is that different types of expertise may be required for the different trials, and this will usually require a larger group. A second reason is that a group monitoring multiple trials may need to meet more frequently to manage the workload; having a larger membership provides some flexibility to committee members in terms of having to miss an occasional meeting (Ellenberg *et al.*, 1993). A third reason is that, with many trials to monitor, it is often efficient to allocate responsibility for 'primary review' of each trial report among the DMC members; a few additional members may be needed to make this a manageable task (see Chapter 6).

When a single group monitors a large number of fairly diverse trials, it will probably be necessary on occasion to invite participation of *ad hoc* members to provide specialized expertise for certain trials. For example, the DMC clinicians serving on the AIDS group mentioned above were either infectious disease specialists or internists with substantial background in treating HIV-infected patients. When a clinical trial addressing ophthalmologic sequelae of HIV infection was undertaken, an ophthalmologist was appointed to participate on the DMC for that particular trial.

DMCs may need to establish procedures to deal with decision-making when some committee members are absent. As an example, in one industry-sponsored trial, a quorum for a seven-member committee was defined as five members including the statistician external to the company sponsor (Williams *et al.*, 1993).

While most DMC recommendations are arrived at by consensus, there are times when consensus cannot be reached and a vote is necessary. For this reason it is sometimes thought useful to constitute the committee with an odd number of voting members. It should be emphasized, however, that recommendations arrived at by vote and not by consensus will not provide as powerful a basis for final decision-making and will inevitably increase the difficulty of this process for the trial organizers/sponsors. Thus, every effort should be made to reach consensus in the recommendations provided to study sponsors and organizers.

3.5 SELECTING THE COMMITTEE CHAIR

The choice of DMC chair is extremely important for the quality of committee functioning. An outstanding chair can optimize the contributions of all members, ensure that all key issues and concerns are dealt with fully and to the group's satisfaction, and can even keep the meeting running on schedule. It is especially important for the committee chair to have the type of broad experience in clinical trials and experience on prior DMCs discussed in the previous section in order for this individual to exert effective leadership of this multidisciplinary group. Because of the special influence the chair may exert, it may be optimal to include at least one other member from the same discipline as the chair.

While all members of the DMC should have the confidence of the other trial components, this is particularly important for the DMC chair. Whatever the appointment mechanism for DMC members (see next section), the selection of the DMC chair should be agreed to by the study sponsor and study chair and/or steering committee representing the trial investigators.

DMC chairs are most often clinicians or statisticians. In Hawkins' (1991) review of 20 DMCs for trials sponsored by the National Eye Institute, 12 were chaired by statisticians and 8 by ophthalmologists; for trials initiated during the 1990s, nearly all chairs have been statisticians (Ferris, personal communication, 1998). DMC chairs for the cardiology, AIDS and cancer trials sponsored by the NIH have generally been clinicians. The particular discipline is less important than prior experience and fundamental leadership ability. In cases where the DMC is not fully independent, it is highly desirable that the DMC chair be one of the independent members, rather than an employee of the government or industry sponsor, or a member of the trial steering committee. Problems can arise (or be perceived to arise) when someone with close ties to the trial serves as chair of the DMC (Strandness, 1995).

3.6 RESPONSIBILITY FOR APPOINTING COMMITTEE MEMBERS

Up to now, this chapter has focused on factors to take into account when selecting members for a DMC. We now address the process of making the selection – and who, in particular, does the selecting.

As noted in Chapter 2, DMCs make recommendations to the study sponsors. Therefore, it is frequently the case that the sponsor appoints the members of the DMC. In doing so, the sponsor (whether government or industry) will often consult with the study leadership, both to solicit suggestions and to ensure that there is not a reason unknown to the sponsor that would make a particular individual an inappropriate choice. This is a highly desirable practice, as the DMC takes on major responsibility to both the sponsor and the study investigators and should have the confidence of both groups.

For industry-sponsored trials, it is usually the case that DMC members are appointed by the pharmaceutical company sponsoring the trial, although some companies may delegate this responsibility to a trial steering committee. In some government-sponsored trials, the DMC appointments are made by the study leadership rather than by the sponsoring agency, although the agency is usually represented in the leadership group. To some extent the appointment process may depend on whether the study is investigator-initiated or agency-initiated.

Policies for NIH trials vary somewhat among institutes and allow for flexibility in arrangements. Geller and Stylianou (1993) reported that some monitoring committees for NIH trials are appointed by the principal investigator for the trial, or the trial steering committee, although in most cases the appointments are made by the sponsoring institute. In cancer cooperative group trials sponsored by the NCI, DMC appointments are made by the cooperative group chair or his/her designee, with concurrence from the Cancer Therapy Evaluation Program of the NCI (Smith *et al.*, 1997). Appointments to DMCs for institute-initiated trials sponsored by the NIAID and by the NHLBI are made by the institutes (Ellenberg *et al.*, 1993; National Heart, Lung and Blood Institute, 2000), although DMCs for investigator-initiated trials may be appointed by the grantee institution. The National Institute for Arthritis and Musculoskeletal Diseases (1999) specifies that DMC appointments should be made by the grantee institution, independently of the principal investigator of the trial. For clinical trials sponsored by the Department of Veterans Affairs (VA), potential DMC members are proposed by the study chair, but final appointments are made by the Chief of the VA's Cooperative Studies Program, 2001).

3.7 REPRESENTATION OF OTHER STUDY COMPONENTS ON THE COMMITTEE

In early implementations of DMCs it was not uncommon to have individuals representing the government sponsor, industry sponsor, study steering committee, and/or the regulatory authority participate fully in DMC meetings and even serve as members. As DMC models have evolved over the past few decades, however, experience has shown that it is almost always advantageous to limit full participation in DMC meetings (i.e., attendance at parts of the meeting in which unblinded safety and efficacy data are considered), and certainly DMC membership, to individuals with no vested interest in the trial conduct or outcome and no primary responsibility for or authority to make changes in the protocol. Representatives of the sponsoring pharmaceutical company clearly have a financial stake in the outcome. Representatives of a government sponsor are less clearly invested in the outcome, but may perceive that more favorable programmatic support will be gained if the trial results go in one direction rather than the other. Both types of sponsors may be confronted with external information that might suggest protocol changes and will be unable to consider such changes objectively if they have knowledge of the impact of the change on the trial results. The major conflict potentially arising with participation of regulatory agency personnel is not with the monitoring of the trial, but later on if that individual is involved in approval decisions for the product under study. Having participated in a consensus-based deliberative process during the course of the trial, the regulator can no longer perform a truly independent review, and may be reluctant to raise concerns that in principle might have been noted (but were not) by the DMC during interim reviews. As for sponsors, steering committee members and trial investigators may have to make decisions as the trial progresses that, if made with knowledge of the interim data, may bias the trial results.

The steering committee member traditionally most likely to participate in all aspects of DMC meetings is the trial statistician. While it is clearly essential to have the data analyzed and presented by someone thoroughly familiar with the study design, the trial statistician (whether employed by an industry sponsor, an academic institution, a clinical research organization or a government agency) may be put in a difficult position by trying to play a dual role as liaison to the DMC and also as a steering committee member. For example, if the DMC recommends a trial modification, the statistician will be aware of the implication of that change on the current data, but other members of the steering committee will not. It will then be problematic for the statistician to participate in an open discussion with other steering committee members concerning whether or not to implement the DMC recommendation. This situation could also create the perception, fair or not, that steering committee decision-making during the trial was driven by the interim data, casting doubt on the integrity of the trial. While we recognize that it has been very common, even typical, for the trial statistician to participate in all DMC deliberations, the potential concerns raised by this model merit serious consideration by trial organizers. This issue is discussed more fully in Chapter 7, where an alternative model is proposed and discussed.

In our collective experience, problems are most likely to occur when individuals representing an industry sponsor participate in the monitoring process. This may be especially true for small companies, when the entire future of the company may depend on the outcome of the trial. We have not become aware of many major problems resulting from the participation of government sponsors, regulators, or steering committee members, but there have been some, as noted in Chapter 5.

The implications of representatives of the various trial components serving as members of DMCs and/or having access to the interim data analyses will be discussed more fully in Chapter 4. As noted in Chapter 6, however, it is valuable to provide an opportunity for such individuals to discuss trial issues with DMC members during open sessions of DMC meetings.

3.8 PREPARATION FOR SERVICE ON A COMMITTEE

With the increase in recognition of the value of an independent DMC to the conduct and interpretation of a clinical trial, the number of trials with such DMCs is correspondingly increasing, and the demand for DMC members is outstripping the supply of individuals with DMC experience. Of course, we cannot in any case limit selection of DMC members to those with prior experience on a DMC; we must, however, consider how best to train prospective DMC members to be constructive, participatory contributors to the monitoring process.

The roles and functions of DMC members are probably best conveyed by direct observation. Thus, a natural source of potential DMC medical and statistical members is the set of individuals with experience in carrying out clinical trials who, by virtue of their role in these trials, have engaged in interaction with trial DMCs. Such individuals may include statisticians preparing and presenting interim analyses for DMC review, or clinician study chairs who may discuss aspects of the trial with the DMC during open sessions (see Chapter 6).

A more prospective approach to 'training by observation' would be to permit one or two DMC members in a trial to bring an 'apprentice' to DMC meetings as an observer. The apprentice would most likely be a more junior person in the member's department who has had fairly extensive clinical trials experience but who has not yet participated in a DMC. Such an arrangement would allow the apprentice all the experience of a DMC member, without assigned responsibility. The apprentice, of course, would be bound by the same confidentiality and conflict-of-interest considerations as the members. We are not aware of any trials organization in which this concept has been implemented, but we believe it could prove to be a highly effective way to prepare potential DMC members for such service. One potential difficulty would be the need for additional funding to support the attendance of apprentices at committee meetings.

More traditional training approaches could also be adopted. Training courses, run over a period of 1–3 days, developed and taught by individuals with extensive experience of serving on DMCs, might be valuable. Such courses could be offered in conjunction with annual meetings of professional societies whose members participate in clinical trials. Material on DMCs might profitably be added to the curricula of general courses on clinical trials methodology offered in schools of medicine and public health. Individuals serving on DMCs that have struggled with particularly difficult problems who then write this DMC experience as a case study and publish it in a journal are also contributing to the training of future DMC members. A number of such papers have appeared in the literature, more frequently in recent years (Armitage, 1999a, 1999b; Bergsjo *et al.*, 1998; Brocklehurst *et al.*, 2000; Cairns *et al.*, 1991; DeMets *et al.*, 1982, 1984; Fleming *et al.*, 1995; Henderson *et al.*, 1995; Pawitan and Hallstrom, 1990; Peduzzi, 1991; Simberkoff, 1993).

REFERENCES

Armitage P on behalf of the Concorde and Alpha Data and Safety Monitoring Committee (1999a) Data and safety monitoring in the Concorde and Alpha Trials. *Controlled Clinical Trials* **20**: 207–228.

Armitage P on behalf of the Delta Data and Safety Monitoring Committee (1999b) Data and safety monitoring in the Delta Trial. *Controlled Clinical Trials* **20**: 229–241.

Bergsjo P, Breart G, Morabia A (1998) Monitoring data and safety in the WHO Antenatal Care Trial. *Paediatric Perinatal Epidemiology* **12** (Suppl. 2): 156–164.

Brocklehurst P, Elbourne D, Alfirevic Z (2000) Role of external evidence in monitoring clinical trials: experience from a perinatal trial. *British Medical Journal* **320**: 995–998.

Cairns J, Cohen L, Colton T *et al.* (1991) Issues in the early termination of the aspirin component of the Physicians' Health Study. *Annals of Epidemiology* **1**: 395–405.

Cooperative Studies Program (2001) *Guidelines for the Planning and Conduct of Cooperative Studies*. Office of Research and Development, Department of Veterans Affairs. http://www.va.org/resdev.

Coronary Drug Project Research Group (1981) Practical aspects of decision-making in clinical trials: the Coronary Drug Project as a case study. *Controlled Clinical Trials* **1**: 363–376.

DeMets DL, Williams GW, Brown BW Jr and the NOTT Research Group (1982) A case report of data monitoring experience: the Nocturnal Oxygen Therapy Trial. *Controlled Clinical Trials* **3**: 113–124.

DeMets DL, Hardy R, Friedman LM, Lan KKG (1984) Statistical aspects of early termination in the Beta-Blocker Heart Attack Trial. *Controlled Clinical Trials* **5**: 362–372.

Ellenberg SS, Myers MW, Blackwelder WC, Hoth DF (1993) The use of external monitoring committees in clinical trials of the National Institute of Allergy and Infectious Diseases. *Statistics in Medicine* **12**: 461–467.

Fleming TR, Neaton JD, Goldman A *et al.* (1995) Insights from monitoring the CPCRA didanosine/zalcitabine trial. *Journal of Acquired Immune Deficiency Syndrome and Human Retrovirology* **10** (Suppl. 2): S9–18.

Geller NL, Stylianou M (1993) Practical issues in data monitoring of clinical trials: summary of responses to a questionnaire at NIH. *Statistics in Medicine* **12**: 543–551.

Hawkins B (1991) Data monitoring committees for multicenter trials sponsored by the National Institutes of Health. I. Roles and membership of data monitoring committees sponsored by the National Eye Institute. *Controlled Clinical Trials* **12**, 424–437.

Henderson WG, Moritz T, Goldman S, Copeland J, Sethi G (1995) Use of cumulative meta-analysis in the design, monitoring and final analysis of a clinical trial: a case study. *Controlled Clinical Trials* **16**: 331–341.

National Heart, Lung and Blood Institute (2000) *Establishing Data and Safety Monitoring Boards (DSMBs) and Observational Study Monitoring Boards (OSMBs)*. http://www.nhlbi.nih.gov/funding/policies/dsmb_est.htm.

National Institute of Arthritis and Musculoskeletal and Skin Diseases (1999) *Data and Safety Monitoring Guidelines for Investigator-Initiated Trials.* http://www.niams.nih.gov/rtac/funding/grants/datasafe.htm.
National Institutes of Health (1998) NIH policy for data and safety monitoring. *NIH Guide*, June 10. http://grants.nih.gov/grants/guide/notice-files/not98-084.html.
Parmar MKB, Machin D (1993) Monitoring clinical trials: experience of, and proposals under consideration by, the cancer therapy committee of the British Medical Research Council. *Statistics in Medicine* **12**: 497–504.
Pawitan Y, Hallstrom A (1990) Statistical interim monitoring of the Cardiac Arrhythmia Suppression Trial. *Statistics in Medicine* **9**: 1081–1090.
Peduzzi P (1991) Termination of the Department of Veterans Affairs Cooperative Study of steroid therapy for systemic sepsis. *Controlled Clinical Trials* **12**: 395–407.
Simberkoff MS, Hartigan PM, Hamilton JD *et al.* (1993) Ethical dilemmas in continuing a zidovudine trial after early termination of similar trials. *Controlled Clinical Trials* **14**: 6–18.
Smith MA, Ungerleider RS, Korn EL, Rubinstein L, Simon R (1997) Role of independent data-monitoring committees in randomized clinical trials sponsored by the National Cancer Institute. *Journal of Clinical Oncology* **15**: 2736–2743.
Strandness DE Jr (1995) What you did not know about the North American Symptomatic Carotid Endarterectomy Trial. *Journal of Vascular Surgery* **21**: 163–165.
Williams GW, Davis RL, Getson AJ *et al.* (1993) Monitoring of clinical trials and interim analyses from a drug sponsor's point of view. *Statistics in Medicine* **12**: 481–492.
Wittes J (1993) Behind closed doors: the data monitoring board in randomized clinical trials. *Statistics in Medicine* **12**: 419–424.

4

Independence of the Data Monitoring Committee: Avoiding Conflicts of Interest

Key Points

- Individuals with important conflicts of interest in regard to a particular clinical trial should not serve on a DMC for that trial.
- The most obvious conflicts are financial, but there can also be intellectual and emotional conflicts of interest.
- Complete elimination of all real, potential and perceived conflicts of interest is generally not possible if one wishes to include DMC members who are knowledgeable and experienced in the medical area being studied.

4.1 INTRODUCTION

The term 'conflict of interest' arises in virtually every setting in which individuals are invited to provide advice, and where that advice might have important and lasting consequences. Conflicts of interest exist naturally as part of normal professional lives. While the occurrence of such conflicts does not necessarily reflect intrinsically inappropriate activity, it would be inappropriate if these conflicts were not acknowledged and dealt with in a proper manner. Judges must recuse themselves from court cases in which they might be swayed by factors other than those presented by the attorneys – for example, personal involvement with a defendant or plaintiff. Members of civic boards faced with decisions on zoning, highway construction, or other similar issues, will not be permitted to participate when they have a personal financial interest in the outcome.

In the research arena, members of NIH study sections do not participate in evaluating and ranking projects submitted by institutional colleagues, family members, or others with whom they may have strong academic or business ties. These conflict-of-interest considerations are also of great importance for members of DMCs.

4.2 RATIONALE FOR INDEPENDENCE

DMC conflicts of interest, whether real or perceived, could arise in several ways that could adversely impact the reliability and credibility of trial results. For example, financial or professional incentives might bias the recommendations of the DMC about trial continuation. In order to capitalize on an apparent early benefit in efficacy, early termination of the trial could be advocated by a product manufacturer wishing to maximize the likelihood of and timing of regulatory approval, or early release of results could be advocated by a lead investigator seeking to benefit professionally from the ability to publish these results in a prestigious journal. With an emerging unfavorable trend, a sponsor or investigator with financial interests may wish to stop prematurely to reduce development expenditures for a new drug, or to avoid demonstrating a drug in current use to be inferior to another product, protecting the future of the drug. If positive results about the benefit-to-risk profile of the product are provided by related trials and are under review by regulatory authorities, a product manufacturer may wish to delay release of a trial's unfavorable interim results until after regulatory action is completed. Conflict of interest could also arise for investigators serving on the DMC of the trial in which they are entering and treating patients. If these investigators change their approach to recruitment and patient care on the basis of early data trends, such action could hamper the ability of the trial to obtain a reliable assessment of the treatments under study.

These types of associations and involvement can, in many cases, complicate the task of providing an independent, fair and rigorous evaluation of the accumulating data and the trial's progress. The expert judgment of the DMC often forms the basis of major decisions about the ongoing conduct of the trial. If DMC members were in a position to benefit – financially, professionally or otherwise – from a particular study result, their expert judgment might be affected by this conflict of interest. Thus, it is essential to ensure as far as possible that this judgment is not unduly influenced by factors other than the needs to safeguard the interests of study participants and to preserve the integrity and credibility of the trial.

Most guidelines for clinical trials, beginning with the Greenberg Report (Heart Special Project Committee, 1988), recommend that individuals with direct financial, intellectual or other conflicts of interest with a drug, device or procedure should not participate in a DMC evaluating that product. These guidelines call for the DMC to be 'independent' of the product manufacturer and the study investigators. Such independence is advocated by the clinical trials guidelines of the National Institutes of Health (1998), as well as the guidelines of the individual institutes (National Eye Institute, 2001; National Heart, Lung, and Blood Institute, 2000a, 2000b; National Institute of Allergy and Infectious Diseases, 2001; National Institute of Arthritis and Musculoskeletal and Skin Diseases, 2000; National Institute of Child Health and Human Development, 2001; National Institute of Diabetes and Digestive and Kidney Diseases, 2001; National Institute on Drug Abuse, 2000; National Institute of Mental Health, 2001) and other agencies such as the Department of Veterans Affairs (Cooperative Studies Program, 2001). This independence is intended to ensure a maximally objective and unbiased assessment of the trial progress and the accumulating safety and efficacy data. Decisions regarding corrective protocol changes or early termination due to safety concerns or established efficacy are often complex and will be much more difficult if some members of the DMC have serious conflicts of interest. Given the special role and responsibility of the DMC, the goal of independence for DMC members is not merely an ideal but is usually an essential.

There are three aspects of independence that we shall explore in more detail. These are: financial issues for both sponsors and investigators; intellectual investment by sponsors, investigators and regulators; and emotional investment by investigators and patient representatives.

4.3 FINANCIAL INDEPENDENCE

Among potential sources of conflict of interest for those involved in monitoring clinical trials, those relating to financial gain usually provide the most obvious concerns. While it is optimal for the DMC to be financially independent of those sponsoring and conducting the trial, achieving this goal is not as straightforward as it may seem.

4.3.1 Sponsors

Pharmaceutical companies typically have important financial conflict of interest pertaining to the conduct of the clinical trial. Ensuring that trial results reflect as favorably as possible on company products could maximize company financial gains. In turn, such maximization would enable the company to provide the highest benefit to its stockholders and employees. Those employees who are primarily responsible for achieving such financial gains (i.e., those who led the drug development effort that resulted in product approval) may accrue particularly large benefits, such as bonuses, salary increases or promotions. Thus, any judgment concerning trial conduct, especially regarding early stopping, that is made by an unblinded company employee engaging in the monitoring process will inevitably be regarded as being potentially biased.

An additional concern regarding sponsor participation in a DMC relates to stock purchases. Employees often wish to invest in their own company as part of their retirement program or just for financial growth. However, individuals cannot legally buy or sell stock if they have information that is considered to be material. A common test of materiality is whether 'Joe Investor's' decision to buy or sell stock would be informed by such information. Companies generally have in place mechanisms to alert individuals to the inappropriate use of such information in making trading decisions, and the Securities and Exchange Commission regularly tracks transactions made by company employees to keep the playing field of trading level. It is important to recognize that some information is so highly technical or subject to interpretation that 'Joe Investor' could not make an educated guess about the meaning of such information. In such an example, information, though privileged, may not be material. Alternatively, access to accumulating efficacy and safety data during the conduct of a pivotal trial could be viewed as material, particularly if the results of the trial will affect the valuation of a company as perceived by outsiders. Thus membership on the DMC and corresponding access to interim data usually would limit the range of acceptable stock activities in which employees could engage during the course of the trial and until the trial results were unblinded.

4.3.2 Academic investigators

Considerations about financial conflict of interest are also very important for academic scientists serving on DMCs. Like sponsors, many academic investigators invest in the stock market as part of their retirement or general financial portfolio. If an investigator serving on a DMC also owns stock in a company sponsoring a trial or in a company having significant competing interests, the 'success' of the product being studied would directly influence the financial interests of the academic investigator, creating a clear conflict of interest. Many view ownership of stock in companies sponsoring the trial to be in conflict with the very role of the trial investigator, even when the investigator does not have access to interim data through DMC membership. Some trials (Healy *et al.*, 1989) have required that all participating investigators divest themselves of any financial investments in the companies that produce either the therapies being studied or directly competitive products.

Determination of potential financial conflicts can usually be made if potential DMC members disclose their investments and forgo participation when a conflict is clear. It may be difficult to identify all potential conflicts, however. For example, the increasing number of mergers and the incorporation of small companies into larger ones can make it a challenge for someone not following industry news closely to recognize which companies would experience significant financial impact from the results in a given trial. While such lack of recognition would preclude occurrence of true conflict of interest for that DMC member, the perception of conflict of interest could harm the ultimate credibility of the trial.

Academic investigators are also subject to other types of financial incentives not directly related to investment interests. Most investigators in large academic or research centers depend on external funding to support their basic science or clinical research laboratories. While federal funding agencies have traditionally been the major sources of such funding, pharmaceutical industry support of academic research has been increasing. Obtaining funding is a very competitive process and often depends on the investigator's connections or familiarity with particular sponsors. Whether the sponsors are federal or private, this dependence for funding could, in principle, create a conflict for any member of a DMC.

Research grants or contracts are generally administered through the institution for which the investigator works, and so the direct financial aspects are kept in proper order and out of the immediate control of the researcher. The impact of such sponsor support cannot be eliminated even in this situation, however. Investigators with large research grants or contracts from an industry sponsor may have a desire to please the sponsor and thereby enhance opportunities for future collaborations. They might unconsciously give undue consideration to the sponsor's interests relative to the interests of the patients and the scientific integrity of the study. Industry sponsors also typically pay investigators a substantial honorarium for their efforts. In some cases, investigators may also be offered all-expense-paid trips to a resort-type location where the consulting takes place. While these payments and rewards are not necessarily inappropriate, they could influence, or at least appear to influence, DMC assessments or recommendations when these investigators serve on the DMC for a trial studying a product made by a company with whom they have such ties.

Similar conflicts could exist for academic investigators who depend heavily on government agencies such as the National Institutes of Health for their research support. Federal sponsors, like some industry sponsors, might expect (or might be perceived to expect) investigators dependent on them for funding to look out for the sponsor's interests while being members of a DMC for a trial they sponsor. Academic researchers, through grant review and other major committee appointments, can become heavily involved in the activities of a federal agency. Even though government agencies typically do not pay a large financial remuneration for such consulting services, these high-profile activities may diminish to some extent that researcher's independence when serving as a DMC member for a trial sponsored by that agency. At the other end of the seniority spectrum, junior academic researchers may be very intent on making positive impressions by taking actions that would please the federal funding agency. Federal as well as industry-sponsored investigators, in considering a recommendation to terminate a trial early for benefit, lack of benefit, or harm, might be influenced by the recognition that early termination could result in decreased funding for that project.

Sometimes financial conflicts can be fairly subtle. A colleague has described her experience as a member of a DMC for a trial being conducted by a small company with limited experience in mounting trials and also limited resources. She agreed somewhat reluctantly to perform and present the interim analyses to the rest of the committee, as well as serve as a member of that committee. This combination of roles very quickly led to recognized conflicts. With her 'DMC hat' on, she had an interest in seeing a variety of supplemental analyses that might (or might not) contribute to a fuller understanding of the emerging results. But each additional analysis she performed meant a larger payment to her from the company. As a result, she felt she was less aggressive in her analytical approach because she was concerned that these additional analyses might be perceived as being motivated by a wish to increase her compensation rather than purely scientific purposes (Wittes, personal communication, 2001). Such conflicts could also occur when different individuals within the same organization played these roles; for example, if a DMC statistician worked in the same academic department or consulting firm that had been contracted to do the statistical analysis and/or data management for the study.

Government agencies, such as the NIH, typically pay a modest honorarium to individuals for services to the DMC. Industry sponsors of clinical trials typically pay members of the DMC much higher honoraria, comparable to the amount a member would receive for consulting or speaking engagements. In principle, any payment or honorarium could be viewed as introducing a conflict. It would be unreasonable, however, to expect those who are asked to serve on a DMC to take time away from their offices and laboratories on weekdays or from families and friends on weekends if there were no incentives – either the reward of public service, in the case of government-sponsored trials, or financial gains, for industry-sponsored trials. Thus, the matter becomes one of amount or degree of the financial incentive. If the level of honorarium for a DMC member is consistent with that individual's standard consulting rate, then the appearance of a conflict is lessened (although not completely eliminated). In contrast, if DMC remuneration were well above the individual's standard rate, the possibility of a conflict would be of much more concern. Honoraria should be expected to cover required preparation time, as well as the meeting day itself. (Honoraria should not be in the form of stock options, for reasons discussed earlier.)

Remuneration rates for DMC activities may very across sponsors and across trials and, as just noted, may vary according to the standard consulting rates of particular members (although in some cases all members may receive the same remuneration). In many trials, a DMC chair or a designated member is required to spend considerably more time and/or to assume considerably greater responsibility for special tasks, such as more frequent assessment of safety data. Thus it may be appropriate to compensate that individual accordingly. Honoraria may be paid on a per-hour or a per-meeting basis, or on an annual basis regardless of the number of meetings. The latter approach has the advantage that DMC members would not have financial incentives to recommend meetings be held more frequently than necessary; it has the disadvantage that DMC members may be reluctant to review data as frequently as might be warranted. Our experience to date, however, has been that such issues have little influence on DMC members, although this would admittedly be difficult to document. In general, the most significant motivations for DMC members to participate arise from their commitment to the science, to their fellow investigators and, most importantly, to the participants in the trial and public health interests.

An extreme case of financial conflict arose in a setting where an individual was the inventor of a cardiac device and had a major financial investment in its manufacture and distribution. This individual also was the primary investigator in a study evaluating its potential use and effectiveness as well as the primary person monitoring the data. As trends began to emerge, the inventor–investor–investigator asked for external experts to advise him on the proper course of action. He was interested in presenting these early favorable but still non-definitive results in order to keep financial and consumer interest in the product at a high level and to maintain a competitive edge in the field. While the external group of experts recommended not presenting or publishing these early results to avoid the chance of releasing false conclusions as well as to reduce the possibility of completing the study in a biased fashion, the individual could not resist the financial pressures. He did present the early results. The *ad hoc* group of experts resigned and sent a letter to the investigator's institution, objecting to this course of action. Other problems involving institutional review and informed consent approval also emerged and the research project was ultimately terminated. This example illustrates the problems with an investigator, who is invested intellectually and financially with an investigational product, participating in the interim monitoring of a clinical trial evaluating that product. In this case, the investigator, the institution, the patients and the research project as a whole would have been better served by having an independent DMC from the beginning to review interim results and to reveal them to the sponsor only when they became scientifically definitive.

Clearly, there are many sources of financial conflicts that could affect the independence of a DMC. Ideally, a DMC should remain free of all such conflicts, although this is not fully achievable given that most DMC members are active researchers and dependent on funding from industry or federal sponsors. In order to maintain the integrity of the DMC process and the clinical trial, individuals having major financial conflicts should be excluded from participation; lesser conflicts may be dealt with by disclosure of financial interests such as sources of research support and consulting activity. Such disclosure reports should be provided annually by DMC members. The DMC chair, the study sponsor or the steering committee of the trial should review these reports to ensure that existing levels of conflict are acceptable. To facilitate this procedure, some trials have developed conflict-of-interest forms, usually focusing on financial conflicts, that researchers complete and update annually. While review of the disclosed activity does not guarantee prevention of all conflicts, it can eliminate many and provide the investigators, the scientific community and the public with a greater degree of confidence regarding the independence and unbiasedness of the monitoring process.

4.4 INTELLECTUAL INDEPENDENCE

While financial conflicts of interest typically cause the most public and professional concern, intellectual conflicts may also exist and in some cases may produce very significant levels of conflict. The DMC collectively should have a reasonably neutral or open position regarding the potential benefits and harms of any experimental intervention. Any DMC member who has a large intellectual investment or a strong opinion for or against the intervention may be less able to review trial progress and the accumulating data objectively. This lack of objectivity could adversely affect the DMC deliberations. DMC members should of course be well informed and knowledgeable, but should not be viewed as crusaders for one point of view. Thus, the goal should be to choose DMC members who are independent intellectually and likely to remain so during the course of the trial.

By the time an intervention such as a drug, device or procedure reaches the stage of a phase III pivotal study, a research team has already invested a great deal of time and energy in the product or procedure. Industry or government sponsors clearly have high hopes that the intervention will work, apart from financial or funding support issues. Careers at the sponsor or an academic institution may have been built on the basic science or concepts behind the intervention. With that level of intellectual investment, an individual or sponsor may have difficulty in critiquing objectively the success or failure of an intervention under investigation in a trial.

Physicians caring for patients with a critical or chronic disease also have intellectual and professional conflicts. They are typically anxious to use new and better interventions for their patients. This desire can sometimes lead them to become interested in (and begin using, if available) a new intervention before it has been definitely established to be safe and effective. Physicians or other basic scientists may also have been involved in pre-phase III studies with an intervention and as a result may be intellectually invested. Again, this level of intellectual investment may reduce the ability of these investigators to maintain a position of neutrality or independence, thus rendering them to be non-ideal candidates to serve on a DMC for a trial evaluating that intervention.

Intellectual investment of DMC members could be acceptable if advocates and skeptics for an experimental intervention have balanced representation on the committee. The Intermittent Positive Pressure Breathing (IPPB) Trial provides an illustration (IPPB Trial Group, 1983). Intermittent Positive Pressure Breathing (IPPB) was a procedure once used to treat patients with chronic obstructive pulmonary disease. It involved a device that delivered a bronchodilator deep into the lung, using forced air. This expensive procedure became popular and part of the standard of care in many places, prior to being evaluated in any definitive trials. The IPPB Trial was designed to compare the clinical effectiveness of this procedure relative to a simple and inexpensive hand-held nebulizer. After several years of patient follow-up, the trial demonstrated that the IPPB procedure had no clinical advantage over the hand-held nebulizer.

One of the DMC members was a prominent pulmonary physician and investigator, experienced in using the IPPB device and procedure but not participating in the trial. As the trial progressed and beneficial trends failed to emerge, this DMC member had to struggle to accept these results that strongly contradicted his own belief and clinical assessment. However, his presence provided a valuable contribution to the DMC and the IPPB Trial due to his strength of character and personal integrity. He was able to articulate what the criticism might be about the trial design and conduct so that these factors could enter into the DMC discussion. Those pulmonary physicians who supported IPPB had the reassurance that their point of view would be presented. While he served as a DMC 'representative' for the many pulmonary physicians who used and promoted IPPB, his point of view was balanced by other DMC members who were somewhat skeptical of IPPB's effectiveness.

Members of regulatory agencies who serve on DMCs can sometimes face intellectual conflicts, although these more likely would affect their regulatory work than their DMC participation. If a regulatory agency representative participated in a DMC that terminated a trial early for benefit, it could be difficult for that individual, who might feel invested in the early termination decision, to then provide a truly objective regulatory review of that product. If a regulatory agency had given approval for a product based on an intermediate outcome or surrogate, they might be biased or in conflict if a subsequent trial had adverse or negative trends, suggesting the need to reverse their prior decision. To avoid these types of intellectual conflict, regulatory agency staff should not be members of a DMC for a trial of a product under their purview.

The use of DMCs has grown rapidly, especially with the increase in phase III trials sponsored by industry. However, the number of clinical trials experts with DMC experience has not increased at a comparable rate. Thus, those people having extensive clinical trials and DMC experience frequently are asked to serve in multiple roles simultaneously. It is natural for those individuals with the greatest expertise to be sought after both as investigators and as DMC members.

Careful thought does need to be given to potential conflicts arising from these multiple roles. It may not be advisable, for example, for a clinical researcher to be an investigator in one trial and be a member of a DMC for a competing trial (i.e., a trial of the same or a similar product in the same patient population). The knowledge gained through the DMC of one trial could impact or unbalance the equipoise necessary to participate as a clinical investigator in the competing clinical trial.

A related issue is the service of an individual on two different DMCs for trials of the same or similar products that are ongoing simultaneously. Although overlapping DMC membership does reduce the independence of the trials to some degree, there can also be important advantages resulting from the enhanced insights provided to the DMC. An illustration of the benefit of having overlapping DMC membership on concurrent related trials was provided by two identically designed trials, one in Europe (EU) and a second in North America (NA), evaluating an investigational therapy in patients with secondary-progressive multiple sclerosis. Both studies were of 4 years' duration. The same individual, a highly regarded neurological scientist, served as DMC chair for both trials. When the NA trial was 2 years from completion, the EU trial was stopped due to significant evidence of efficacy. In spite of strong public pressure to stop the NA trial, its DMC recommended continuation after the committee completed a careful review of the final results from the EU trial and interim results from the NA trial. The sponsor, clinical investigators and study participants, who remained blinded to results from the NA trial, were more willing to accept this recommendation to continue the NA trial due to their trust in the objectivity and judgment of the DMC chair, and due to their recognition that he fully understood the strength of evidence provided by the EU trial. The NA trial was successfully completed on schedule. Its final results contradicted the EU trial results, raising doubts about the product's effectiveness in that clinical setting.

Neither DMCs nor individual members should be co-authors of the primary trial publications. If the DMC were to anticipate authorship on the primary publication, it would no longer be an independent entity. It is appropriate, however, for the DMC to be acknowledged, along with all other committees necessary to conduct a trial. The acknowledgement of an independent DMC, in addition to crediting the effort of the committee, adds credibility to the trial conduct and results.

4.5 EMOTIONAL CONFLICTS

We have already discussed how a physician's desire to offer a patient a new intervention, with the hope it would be better than current therapy, could produce emotional as well as intellectual conflicts. Another potential entanglement arises when DMC members have special professional relationships with study leaders or sponsor representatives. A senior faculty member might be appointed to a DMC for a trial chaired by a former student or a medical resident. Concerns for career issues relating to the colleague may also influence the DMC member's perspective and create an obstacle to a fully independent review. For example, terminating a trial early due to lack of effect would likely result in a loss of funding for the trial investigators. These types of conflicts are sometimes difficult to determine in advance and therefore avoid. However, given that those who participate in clinical trials in a specific field often know each other and may have trained with each other, DMC members should be conscious of this potential conflict. NIH study sections, which review grant applications, typically avoid such conflicts by asking members to recuse themselves when considering any grant for which the grant reviewer has a close professional relationship to the applicant. The same sensitivity should apply in the setting of a DMC.

Another constituency increasingly involved in clinical trials is the patient advocacy group. Patient advocacy groups, such as those for AIDS or breast cancer, have worked to increase research funding and to speed up the availability of newer therapies. These groups have been very successful in encouraging the federal government and private industry to allocate additional resources for further research. The commitment of many such advocacy groups to their cause could, however, make it difficult for their members to remain independent, as was noted in Chapter 3. For example, an early favorable trend could trigger an emotional appeal to terminate early or to release interim results, while the purely scientific considerations might indicate a need for more definitive data. Such an appeal at a DMC meeting may be awkward, potentially disruptive to deliberations, and not in the best interests of study participants or investigators. However, an individual who represents patients' interests, and yet is not invested directly in the sponsor, the product or advocacy for the specific trial in question, may contribute positively as a DMC member if that individual can participate dispassionately and abide by the constraints of confidentiality.

4.6 INDIVIDUALS WITHOUT CONFLICTS

As noted earlier, avoiding all potential or existing conflicts can be difficult if not impossible. Individuals who have relevant expertise and experience gain that status because they are involved in clinical research. Thus, they have their own sources of research support and research agendas. The DMC, the trial investigators and the trial participants would not be served well by DMC members who had no experience in the clinical area under investigation. The challenge in recruiting DMC members is to minimize as many sources of conflict as possible, recognizing that some may be impossible to totally eliminate.

It is necessary for an effective DMC to have members with a diversity of expertise, experience and opinions. If the obvious and important conflicts of interest are eliminated, and others are identified through disclosure, then the appointment process and the DMC monitoring activities will have an excellent chance of being successful. In these circumstances, most DMC members are able to set aside any lesser conflicts to best serve the study participants, study investigators, the research community and the study sponsors. Our collective experience has been very favorable in this regard.

REFERENCES

Cooperative Studies Program (2001) *Guidelines for the Planning and Conduct of Cooperative Studies.* Office of Research and Development, Department of Veterans Affairs. http://www.va.org/resdev.

Healy B, Campeau L, Gray R *et al.* (1989) Conflict-of-interest guidelines for a multicenter clinical trial of treatment after coronary artery bypass-graft surgery. *New England Journal of Medicine* **320**: 949–951.

Heart Special Project Committee (1988) Organization, review and administration of cooperative studies: A report from the Heart Special Project Committee to the National Advisory Heart Council, May 1967. *Controlled Clinical Trials* **9**: 137–148.

Intermittent Positive Pressure Breathing Trial Group (1983) Intermittent positive pressure breathing therapy of chronic obstructive pulmonary disease–a clinical trial. *Annals of Internal Medicine* **99**(5): 612–620.

National Eye Institute (2001) *National Eye Institute Guidelines for Data and Safety Monitoring of Clinical Trials*. http://www.nei.nih.gov/funding/policy/policy6.htm.

National Heart, Lung, and Blood Institute (2000a) *Responsibilities of Data and Safety Monitoring Boards (DSMBs) Appointed by the NHLBI*. http://www.nhlbi.nih.gov/funding/policies/dsmb_inst.htm.

National Heart, Lung, and Blood Institute (2000b) *Responsibilities of Data and Safety Monitoring Boards (DSMBs) Appointed by Participating Institutions*. http://www.nhlbi.nih.gov/funding/policies/dsmb_othr.htm.

National Institute of Allergy and Infectious Diseases (2001) *Guide to Requirements for Research Grants Involving Human Subjects*. http://www.niaid.nih.gov/ncn/tools/humansubjects/guidereq.htm.

National Institute of Arthritis and Musculoskeletal and Skin Diseases (2000) *Data and Safety Monitoring Guidelines for Investigator-Initiated Clinical Trials*. http://www.niams.nih.gov/rtac/funding/grants/datasafe.htm.

National Institute of Child Health and Human Development (2001) *NICHD Policy for Data and Safety Monitoring*. http://www.nichd.nih.gov/funding/datasafety.htm.

National Institute of Diabetes and Digestive and Kidney Diseases (2001) *NIDDK Data and Safety Monitoring Guidelines For Clinical Trials*. http://www.niddk.nih.gov/patient/clinical_trials/niddkgenmonit.pdf.

National Institute of Mental Health (2001) *NIMH Policy on Data and Safety Monitoring in Clinical Trials*. http://www.nimh.nih.gov/research/safetymonitoring.cfm.

National Institute on Drug Abuse (2000) *Data and Safety Monitoring Board Standard Operating Procedures*. http://www.nida.nih.gov/funding/dsmbsop.html.

National Institutes of Health (1998) NIH policy for data and safety monitoring. *NIH Guide*, June 10. http://grants.nih.gov/grants/guide/notice-files/not98-084.html.

5

Confidentiality Issues Relating to the Data Monitoring Committee

Key Points

- Trial integrity is best protected when interim data comparing treatment groups are seen only by DMC members and the statistician preparing the interim reports.
- Separate reports presenting aggregate data on administrative aspects of the trial can be shared with the sponsor and trial leadership.
- In limited circumstances there may be a strong rationale for wider access to comparative interim data.
- The DMC should have access to fully unblinded data, with actual treatments and not just codes available for its review.

5.1 RATIONALE

An important principle guiding the functioning of data monitoring committees is that members of the DMC should ideally be the only individuals (other than the statistician(s) performing the interim analyses) with access to interim data on the relative efficacy and relative safety of treatment regimens. This principle is justified by the need to minimize the risk of widespread prejudgment of unreliable results based on limited data. As discussed in this chapter, this prejudgment could adversely impact rates of patient accrual, continued adherence to trial regimens, and ability to obtain unbiased and complete assessment of trial outcome measures. This prejudgment could also result in publication of early results that might be very inconsistent with final study data on the benefit-to-risk profile of the study interventions.

DeMets *et al.* (1995) discuss the justification for this confidentiality principle:

> Ensuring the confidentiality of interim results during the conduct of a clinical trial is critical. Without an assurance of confidentiality, a DMC cannot fulfill the responsibility with which it is entrusted. Interim data on treatment benefit and safety may not be mature and usually are not scientifically convincing. If these data are leaked to the investigators, the scientific community, or the public, rumors and inappropriate prejudgments could slow or stop further recruitment and bias patient evaluation, making it impossible to complete a trial or to provide proper scientific evaluation.

An important HIV/AIDS study illustrates how confidentiality enabled successful completion of a trial when interim analyses provided very misleading early results. As discussed in Chapter 1, this trial provided a comparison of two therapies, zalcitabine (ddC) and didanosine (ddI) (Abrams *et al.*, 1994). At an early interim analysis, the patients randomized to the ddI regimen had only half as many primary study events (i.e., symptomatic AIDS events or deaths) as those randomized to ddC (19 vs. 39 events; $p = 0.009$). The patients receiving ddI also had achieved significantly higher CD4 levels ($p = 0.009$), a primary measure of the ability of the regimen to protect the patient's immune system.

In its review of these interim data, guided by the conservative O'Brien–Fleming group sequential monitoring guideline and by extensive consideration of all available data, the DMC judged that these early trial results did not provide reliable evidence about the relative efficacy of these treatments. The trial was continued. At its scheduled completion, when the trial had obtained the protocol-specified fourfold increase in primary study endpoints, the results had changed strikingly. The apparent advantage of the ddI regimen in preventing these primary endpoints had disappeared, and the patients treated with this regimen actually had a higher death rate. DeMets *et al.* (1995) observed:

> Broad dissemination of early trial results (at the interim analysis) would very likely have resulted in widespread prejudgment that ddI had proven to be superior to ddC, foreclosing on the opportunity to obtain the much more reliable and strikingly different later assessments about the relative efficacy of these interventions.

Considerable evidence exists to document the adverse risks to trial integrity resulting from early release of interim efficacy data to those who are not experienced in the monitoring process. For example, in the setting of oncology trials, Green *et al.* (1987) performed a matched analysis of large randomized clinical trials from two major cancer cooperative groups. One of these groups revealed interim results on efficacy only to members of a DMC, while the second did not have a DMC and circulated interim results widely to investigators and others. This analysis provided substantial evidence that DMCs contribute positively to preserving the integrity of prospective clinical trials. In the group without a DMC, 50% of the studies showed declining patient accrual rates over time. In addition, some studies were inappropriately terminated early due to prejudgments and inability to complete accrual, thereby yielding equivocal results. Final results of other completed studies were inconsistent with prematurely published early positive results, leading to inconsistencies in the literature. In contrast, the studies in the group having DMCs were free of these problems.

The benefits provided by DMCs in these oncology trials were achieved even though therapy was usually delivered in an unblinded manner. Even in settings in which investigators are aware of the treatment assignments of their own patients, trials are more likely to be completed successfully when these investigators are blinded to the comparative data from patients managed by other investigators and at other centers. (Of course, the successful blinding of overall results depends on each investigator maintaining the confidentiality of information regarding treatment outcomes in patients from their center.)

The following examples further illustrate the problematic effects in oncology trials of early release of interim efficacy and safety data to non-DMC members.

Example 5.1: Preoperative radiation treatment for rectal cancer

Patients about to undergo surgical treatment for rectal cancer at Toronto's Princess Margaret Hospital were randomized to preoperative radiation treatment versus a control regimen involving surgery alone. The trial was stopped early, with an enrollment of only 125 patients. The investigators reported that early results from the study had shown 'no difference between the two groups' in patient survival (Rider *et al.*, 1977). The authors indicated that the available sample size was much smaller than intended because interim results had been regularly available to all participating clinicians and because 'the absence of any trend in survival during the early years caused the study to die a natural death'.

Wide dissemination of results on relative efficacy of treatment regimens led to an early loss of interest by physicians responsible for patient accrual. Thus, even though the trial was not 'actively' terminated by the lead protocol team, it was 'passively' terminated by prejudgments of accruing physicians, yielding inconclusive results and necessitating the subsequent conduct of a 552-patient confirmatory trial by the Medical Research Council in the United Kingdom (Medical Research Council Working Party, 1984).

Example 5.2: Cancer intergroup study 0035 – fluorouracil plus levamisole in colon cancer

A joint study of the cooperative cancer groups sponsored by the National Cancer Institute involved randomization of 971 stage III colon cancer patients, within 1 month of their clinically complete surgical resection, to adjuvant treatment with levamisole alone or with 5-FU plus levamisole, or to no adjuvant treatment (Moertel *et al.*, 1990). Final analysis was to be performed after 500 deaths had occurred, with interim analyses planned after each group of 125 deaths. The primary intent of the trial was improvement in long-term survival, with a reduction in the rate of disease recurrence providing important supportive information. The O'Brien–Fleming group sequential procedure was used to guide decisions about early termination.

Patient accrual began in March 1984, and was completed in October 1987, prior to the first interim analysis of efficacy results that occurred in spring 1988. At that analysis, the evidence was quite strong that the 5-FU plus levamisole regimen reduced the rate of recurrence of disease. Median follow-up for survival at that analysis was a relatively short 18–24 months, however, with only small trends for survival improvement apparent. Thus, the DMC recommended that the study should continue and remain blinded.

In late summer 1988, the DMC decided the interim results should be shared with a small group of leaders from the FDA and NCI, to facilitate the regulatory review process should the trial be terminated after the second interim analysis. These individuals promised to maintain confidentiality. Nevertheless these confidential relative efficacy results were circulated more widely shortly after the meeting with the DMC, ultimately leading to an editorial in *Science* (Marx, 1989) that challenged the ethics of the DMC's recommendation to continue the trial. Even though accrual to the trial was finished and all patients had completed their 12-month course of chemotherapy by autumn 1988, this violation of confidentiality did present significant risks for prejudgment of unreliable early information that could have adversely impacted the successor trial, a placebo-controlled evaluation of 5-FU plus leucovorin, that was in the midst of its enrollment and treatment period in 1988–89.

In summary, maintaining confidentiality of interim results is indeed of critical importance in the DMC's effort to ensure trial integrity and credibility. This confidentiality minimizes the risk of widespread prejudgment of early unreliable information about efficacy and safety. Such prejudgment could adversely impact the ability to achieve timely accrual of study participants, continued adherence to trial regimens, as well as unbiased and complete assessment of trial outcome measures, not only for the study monitored by the DMC but also for concurrent related trials.

5.2 LIMITS OF CONFIDENTIALITY

Several practical questions relating to confidentiality regularly arise during clinical trials monitored by a DMC. What interim results can be shared beyond the DMC and with whom? How can the DMC fully benefit from the special insights of those who are involved in trial design and conduct and yet must remain blinded? Who should attend DMC meetings other than DMC members? Who should prepare the unblinded reports and serve as the liaison between the DMC and the database? Finally, are there special circumstances in which it will be appropriate to release interim data more widely?

5.2.1 Interim analysis reports

Although preserving confidentiality of comparative interim results is extremely important, it is also important to regularly exchange non-confidential information among various parties who share responsibility for the successful conduct of the trial. This can be accomplished by preparing two types of reports for dissemination and discussion at different sessions within each DMC meeting, one containing administrative and/or aggregate data that can be freely shared ('open' reports) and the other containing confidential data on the relative efficacy and safety of the treatments being compared ('closed' reports). In Chapter 6 we will discuss a format for DMC meetings that provides for these two different kinds of sessions, each focusing on one of the reports. We use the terms 'open' and 'closed' to refer to the sessions and the corresponding reports.

The open reports, to be made available to all who attend the open session of the DMC meeting, generally should include data on recruitment, eligibility violations and baseline characteristics, adherence to interventions, and currency and completeness of follow-up information (see Table 5.1). Most information in the open report is pooled by treatment regimen. The open report should contain no information that is directly or indirectly informative about the efficacy and safety of the study interventions.

Closed reports, available only to those attending the closed session of the meeting, should provide a display by treatment group for all data elements that had been presented only in the aggregate in the open report. In addition, the closed report provides analyses of primary and secondary efficacy endpoints, subgroup and adjusted analyses, and analyses of adverse events and symptom severity.

Table 5.1 Open statistical report: a typical outline

- One-page outline of the study design, possibly with a schema
- Statistical commentary explaining issues presented in open report figures and tables
- DMC monitoring plan and summary of open report data presented at prior DMC meetings
- Major protocol changes
- Information on patient screening
- Study accrual by month and by institution
- Eligibility violations
- Baseline characteristics (pooled by treatment regimen)
  - Demographics
  - Laboratory values and other measurements
  - Previous treatment usage and other similar information
- Days between randomization and initiation of treatment (pooled by treatment regimen)
- Adherence to medication schedule (pooled by treatment regimen)
- Attendance at scheduled visits (pooled by treatment regimen)
- Reporting delays for key events (pooled by treatment regimen)
- Length of follow-up data available (pooled by treatment regimen)
- Participant treatment and study status (pooled by treatment regimen)

Closed reports should also provide analyses of lab values (see Table 5.2). These open and closed reports are optimally prepared by a biostatistician independent of the trial leadership, as described in Chapter 7.

5.2.2 Access to aggregate data on efficacy and safety outcomes

Controversies have arisen regarding the level of confidentiality that must be maintained for aggregate data on efficacy and safety outcomes – that is, pooled data giving the total number of events across all study groups. Since such data often provide suggestive information regarding the relative benefit-to-risk profiles of the treatments being compared, in most settings these data should not be included in the open report. Consider a clinical trial of an experimental drug in advanced cancer patients, where historical evidence indicates the control regimen should yield approximately 15% two-year survival. When one-half of the trial's targeted number of endpoints have occurred, pooled data estimates of two-year survival of 25% or 10% could give a strong impression that the experimental regimen is effective or ineffective, respectively. Even if that impression is incorrect, resulting actions taken by trial investigators, sponsors or patients could compromise trial integrity and credibility. It could also be inappropriate to include pooled data on secondary endpoints in an open report. Consider, for example, a clinical trial designed to assess the effect of a behavioral intervention in preventing transmission of HIV. Release of early data showing a substantial reduction (or no reduction) in the surrogate endpoint of self-reported risk-taking behavior could lead to prejudgment about the efficacy of the intervention, or might lead to a data-driven reformulation of trial primary or secondary endpoints.

The proper level of access to aggregate efficacy and aggregate safety data needs to be determined on a trial-by-trial basis. Inclusion of such data in an open report would be appropriate if this information could broadly inform trial investigators and care-givers about how to enhance the quality of trial conduct, while not

Table 5.2 Closed statistical report: a typical outline

- Detailed statistical commentary explaining issues raised by closed report figures and tables (by coded treatment group, with codes sent to DMC members by a separate mailing)
- DMC monitoring plan and summary of closed report data presented at prior DMC meetings
- Repeat of the open report information, in greater detail by treatment group
- Analyses of primary and secondary efficacy endpoints
- Subgroup analyses and analyses adjusted for baseline characteristics
- Analyses of adverse events and overall safety data
- Analyses of lab values, including basic summaries and longitudinal analyses
- Discontinuation of medications
- Information on crossover patients

providing clues about the relative benefit-to-risk profiles of the treatments being compared. An illustration is provided by the trial of erythropoietin in hemodialysis patients with congestive heart failure (Example 2.8 in Chapter 2). It was clearly known that erythropoietin substantially impacted a biological marker, hematocrit level. The randomized trial was designed to determine whether benefit on a long-term clinical endpoint, patient survival, could be achieved by the intervention, with dosing titrated in a manner to achieve an intended level of effect on the marker. In this instance, data on changes in the biological marker provide important insights about adherence to the study regimen without providing any new insights about efficacy and, hence, would be very appropriately included in the open report. A similar illustration is provided by the ongoing NIAID-sponsored ESPRIT trial that is evaluating the ability of IL-2 to reduce the occurrence of AIDS-defining events, mediated through its previously established immunologic effects represented by changes in the biological marker, CD4 level. Data on changes in CD4 level could be provided in the open report since insights from this information would be limited to quality of adherence to the protocol-specified regimens.

Even when it is not proper for aggregate efficacy and safety data to be widely distributed through inclusion in the open report, such data could be provided to selected individuals who 'need to know' such information to carry out their ethical or scientific responsibilities in the conduct of the trial. In studies with endpoint adjudication committees, for example, the number of endpoints submitted by sites for adjudication will be known at least to the members of that committee. The study chair and/or certain members of the steering committee may also need to have access to that information if they are responsible for monitoring the work of the adjudication committee. As will be discussed in section 5.2.4, a trial's medical monitor who is responsible for providing timely reporting of serious adverse events to regulatory authorities would have at least indirect access to aggregate safety data. In addition, since the determination of the need for sample size adjustments is usually based on the event rate for the primary endpoint in the pooled data, such information would need to be provided to the person charged with this responsibility. In any of these settings in which individuals are provided access to aggregate efficacy and safety data on a 'need to know' basis, these individuals should maintain the confidentiality of this information except where it is necessary to do otherwise to carry out their ethical or scientific responsibilities in the conduct of the trial.

5.2.3 The steering committee and maintaining confidentiality

Many clinical trials will have a steering committee (SC) that will typically be a small multidisciplinary group of individuals who collectively have the scientific, medical and clinical trial management experience to design, conduct and evaluate the clinical trial. The SC usually includes representatives of the sponsors and the principal investigators, possibly supplemented by other clinical scientists with special expertise and/or experience with the issues addressed by the trial. As described in Chapters 3 and 7, the SC should share the responsibility with the DMC for safeguarding the interests of participating patients and for the conduct of the trial.

In order to allow the DMC to have adequate access to insights from the SC, it is optimal for the SC members to either be present at or be provided a telephone link for the open session of the DMC meeting. They would also have full access to data in the open report. However, the SC should not have access to the primary or secondary endpoint efficacy data or safety data from the closed report unless the DMC has recommended early termination of the trial.

Following a DMC recommendation for early termination, the SC would be provided access to the unblinded interim analyses so that they could make an informed judgment about whether the trial should be stopped. When the trial addresses a question that has regulatory implications, the sponsor representative on the SC would ordinarily contact the regulatory authority to ensure that all relevant regulatory issues have been considered. In order to facilitate continuation of the trial in the unlikely event that the SC rejects the DMC recommendation for termination, the SC must maintain confidentiality of all information it receives other than that contained in the open reports until after the trial is completed or until it has made a decision for early termination. If the SC does decide to continue the trial, in addition to maintaining confidentiality of such information, it should also be blinded to subsequent closed report information. Because of the problems regarding confidentiality should the trial continue, the practice in some trials is to initially unblind only the SC chair and a sponsor representative when a DMC recommends early termination. This practice protects confidentiality, but at the cost of a narrower perspective with regard to whether or not to accept the DMC recommendation.

The SC can communicate information in the open report to the sponsor's senior management and to other interested parties. The SC can also inform the trial sponsors of the DMC-recommended alterations to study conduct or early termination in instances in which the SC has reached a decision agreeing with the recommendation. The SC should prepare minutes of its meetings that consider significant recommendations made by the DMC. The content of these minutes, and how they are distributed, will be discussed in Chapter 6.

5.2.4 Settings and procedures allowing broader unblinding

As with efficacy data, comparative data on safety in general should not be released prematurely to the sponsor or to other non-DMC members. Safety information such as numbers of deaths (either by treatment regimen or pooled over treatment regimens) could provide direct or indirect evidence about relative efficacy results, depending on the trial's primary endpoint. More broadly, since the relative safety data can be quite extensive even in the early stages of a trial, inappropriate judgments about the relative benefit-to-risk profiles for two treatments might be made early during the course of the trial. This could compromise the ability to obtain the longer-term efficacy and safety data that would be required to make more reliable assessments about the true benefit-to-risk profiles of the two treatments.

At times, the DMC may be asked to allow broader access to comparative safety data during the conduct of the study. Early release of some information on safety could be potentially important for study participants in certain cases when unexpected serious adverse effects are observed, particularly when a refinement in the dosing/schedule of a study regimen or in the selection criteria for the trial could substantially reduce the risk of that adverse effect. In such circumstances the DMC will have to balance potential risks to patients with the potential risk to the study of release of information. The next example illustrates this situation.

Example 5.3: Hormone replacement therapy in post-menopausal women

The HERS trial evaluated the use of hormone replacement therapy (HRT) in post-menopausal women with a history of cardiac disease. Mortality and cardiovascular morbidity were the primary and secondary outcomes.

At an interim analysis, the DMC noted a statistically significant increase in deep vein thrombosis (DVT) for patients on HRT. The DMC discussed various options, including: continuing the trial with no release of safety information; terminating the trial due to DVT adverse effects; and continuing the trial but informing patients of the DVT risk. Millions of women currently were using HRT in hopes (at least in part) of lowering their risk of cardiovascular morbidity and mortality, yet it remained unknown whether it was effective. If effective, HRT would be a useful intervention even with its associated risk of DVT. After careful deliberation, the DMC recommended that the HERS Steering Committee publish a brief report alerting study participants and HRT users in the general community that women with a cardiovascular history taking HRT were at increased risk of DVT. The DMC also recommended that women in the study who became immobilized, and were thereby at increased risk of DVT, should stop study treatment until they became active again. By reporting these DVT results early, women on HRT were informed about potential risks. The HERS trial continued to show that HRT may increase the risk of DVT early on, but that this risk diminished over time.

Sponsors, investigators or regulators who, during the conduct of the study, are planning future studies, considering future steps in resource allocation or product development, or preparing for regulatory review may perceive the need to see preliminary safety and/or efficacy data. The DMC should have established procedures to evaluate or act on such special requests to provide limited access to evolving study information. When release of data would not unblind comparative treatment efficacy and safety results or jeopardize the successful completion of the trial, the request could readily be granted. Examples of information that might be requested would be event rates by baseline characteristics for participants in the control arm (in trials where access to pooled event rates is only available to the DMC), or rates of adverse events in the control arm. Of course, any information routinely included in open reports typically could be released.

A special case of release of interim data is the 'accelerated approval' process in the USA by which a new product holding substantial promise for treatment of a life-threatening disease may receive marketing approval on the basis of interim data on surrogate endpoints. FDA approval is granted with the understanding that the ongoing trial must be completed and show benefit on important clinical endpoints in order to receive full approval. Substantial risks to trial integrity clearly arise with this practice, most commonly seen in settings such as HIV/AIDS or oncology, since release of information suggesting the superiority of the new treatment, even though not validated by clinical data, may create difficulties in completing the trial and obtaining adequate data on clinical endpoints. Furthermore, if such early release leads to marketing approval while the trial is still ongoing, the power of the trial could be eroded by declining rates of accrual and by diluted estimates of treatment effect resulting from 'dropins' of control patients to the newly approved experimental regimen. Experience has so far indicated that such trials can be successfully completed, but it is well recognized that in enabling earlier access of seriously ill people to promising interventions the risk of licensing agents that could be biologically active yet clinically ineffective is increased.

It is important for the sponsor of the trial to provide regulatory bodies with timely case-by-case reports of serious adverse events. Typically, a sponsor-appointed individual, often called the medical monitor, would be given this responsibility. The medical monitor would obtain immediate access to patient specific information on serious adverse events, blinded to treatment whenever possible. After reviewing such events, the medical monitor would ensure that relevant information is promptly reported to regulatory authorities. Any insights that the medical monitor would obtain regarding aggregate safety data, overall or by intervention group, should not be shared with non-DMC members.

5.2.5 Some illustrations of broader unblinding

Policies and procedures to ensure that the DMC has exclusive access to interim efficacy and safety data should be implemented consistently. Exceptions should be rare, and should require clear justification that the ability to complete the trial, in a manner that would reliably answer the questions it was designed to address, would be fully maintained or even enhanced by allowing some carefully determined and limited level of unblinding.

We begin with an illustration in which limited release of interim results on efficacy was granted. Another illustration will be presented in which confidential information was shared between two DMCs that were monitoring concurrent and identically designed trials.

Example 5.4: Prevention of symptomatic cytomegalovirus disease

The Community Program for Clinical Research in AIDS conducted a placebo-controlled trial, entitled CPCRA 023, evaluating the effect of oral ganciclovir on prevention of symptomatic cytomegalovirus (CMV) retinal and gastrointestinal mucosal disease in HIV-infected patients (Brosgart *et al.*, 1998). The trial was initiated in April 1993. At its midpoint, in July 1994, data were reported from a related trial, entitled Syntex 1654 (Spector *et al.*, 1996). Analyses of that study revealed a 55% reduction in the rate of CMV disease and a nearly significant reduction in mortality (see Table 5.3).

After extensive discussions, the DMC of the CPCRA 023 trial concluded that despite these positive data it would be very important to continue the CPCRA study. Two major considerations justified that conclusion. First, as shown in Table 5.3, the available 023 results suggested only a small effect of oral ganciclovir on prevention of symptomatic CMV disease and the mortality trend was actually in the wrong direction (Fleming *et al.*, in press). Second, because the Syntex trial required bimonthly funduscopic screening exams performed by ophthalmologists, the rate of CMV events in the control arm of that trial was twice the rate observed in the control arm of 023. The 023 DMC was concerned that, in the Syntex trial, ganciclovir might only be reducing the occurrence of asymptomatic cases of CMV disease. Such cases were not being captured in 023 since these were considered to be of limited clinical relevance.

With pronouncements claiming established benefit of ganciclovir following public release of the Syntex results, and given the strong advocacy in the HIV/AIDS community for broad and early access to promising interventions, the DMC recognized that achieving continued compliance to the control regimen during the remaining 12 months of the trial would be difficult. To restore a sense

Table 5.3 CPCRA 023: oral ganciclovir and Prevention of CMV disease

July 94	CPCRA 023		Syntex 1654	
	Ganciclovir	Placebo	Ganciclovir	Placebo
Sample size	646	327	486	239
CMV disease	40	23	76	72
(RR/p)*	(0.87 / 0.60)		(0.45 / 0.0001)	
Death	58	23	109	68
(RR/p)*	(1.27 / 0.34)		(0.71 / 0.052)	

*Relative risk (RR) estimates and p-values obtained from the Cox proportional hazard regression models.

of equipoise within the HIV/AIDS community, the 023 DMC recommended making an immediate limited disclosure of key current results. Letters were sent in August 1994 to the study patients, their physicians and institutional review boards, summarizing the Syntex study results and stating that the 023 results did 'not support the conclusions found in the Syntex study'. These also stated that 'Data from the CPCRA CMV study, at this time, do not show that CMV disease occurs more often in patients taking placebo than in patients taking oral ganciclovir', and 'Data from the CPCRA CMV study, at this time, do not show that patients taking oral ganciclovir live longer than those taking placebo'. After receiving these letters, only a minority of the patients chose to exercise an option to immediately receive open-label oral ganciclovir.

The trial was successfully completed. The final results of 023, obtained in July 1995, are presented in Table 5.4. While the adverse mortality trend disappeared, there was still no evidence of a treatment-induced clinically meaningful reduction of occurrence of symptomatic CMV disease.

This trial illustrates that, while early termination of a given trial might be appropriate when a companion trial reports significant results, continuation could also be justified when current results in the given trial and design differences between the two studies are carefully considered. The trial also illustrates that, in such settings, limited release of key outcome data might be justified when such release could restore a proper sense of clinical equipoise, in turn enhancing the opportunity to obtain needed insights about the risk-to-benefit profile of promising interventions. However, it should be recognized that such circumstances are extremely rare. In fact, this illustration of early release of outcome data to restore clinical equipoise represents the only such occurrence in the experience of the authors of this book.

The next example illustrates sharing of confidential information between two DMCs that are monitoring concurrent related trials. While such sharing is not advocated on a routine basis (see Dixon and Lagakos, 2000), in selected cases it

Table 5.4 Interim and final results in the CPCRA 023 clinical trial

	July 94		July 95	
	Ganciclovir	Placebo	Ganciclovir	Placebo
Sample size	646	327	662	332
CMV Disease	40	23	101	55
(RR/*p*)*	(0.87 / 0.60)		(0.92 / 0.60)	
Death	58	23	222	132
(RR/*p*)*	(1.27 / 0.34)		(0.83 / 0.09)	

*Relative risk (RR) estimates and *p*-values obtained from the Cox proportional hazard regression models.

may greatly enhance the ability of a DMC in its role of safeguarding the interests of trial participants and protecting trial integrity (Armitage, 1999).

Example 5.5: CPCRA 007 – combination antiretroviral therapy in HIV/AIDS

The CPCRA 007 study was initiated in mid-1992 to determine whether duration of survival, free of progression to symptomatic AIDS-defining events, could be improved by either the addition of didanosine or zalcitabine to zidovudine alone (Saravolatz *et al.*, 1996).

Because ddI and ddC were administered in different forms, performing the trial in a double-blind manner would have required the use of multiple placebos. The study organizers wished to avoid having to give ddC placebo to patients randomized to receive ddI, ddI placebo to patients randomized to receive ddC, and more than one placebo to the control group. To achieve this substantial reduction in daily administration of placebo capsules, the final trial design employed a two-level randomization (Figure 5.1). The first level was an unblinded randomization to the ddI group versus the ddC group, followed by a secondary randomization in which two-thirds of the ddI group were to receive active ddI and one-third to receive ddI placebo. Similarly, in the ddC group, two-thirds were randomized to active ddC and one-third to ddC placebo. In this study design, 400 patients received ddI plus zidovudine and 400 patients received ddC plus zidovudine. None of those patients received placebos. In the control arm that included 400 patients who received zidovudine, 50% received the ddI placebo and 50% received the ddC placebo.

Just prior to the mid-point of the trial, in August 1993, the DMC reviewed the results shown in Table 5.5. Twice as many patients on one of the four study arms (arm A) as on another (arm B) (33 vs. 16) had experienced death or progression to symptomatic AIDS-defining events. The nominal p-value for this difference

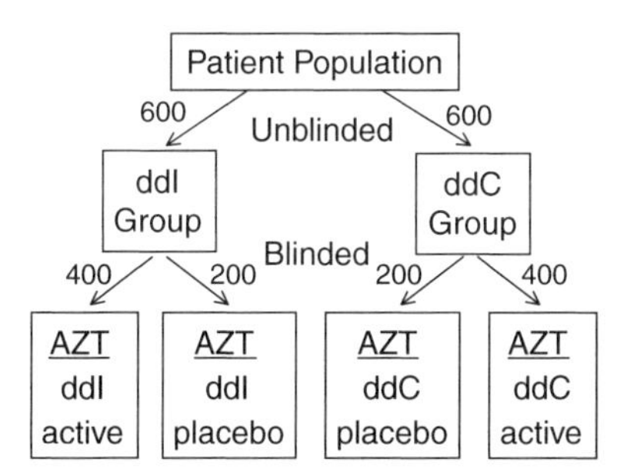

Figure 5.1 Study design for the CPCRA 007 trial. ddI, didanosine; ddC, zalcitabine; AZT, zidovudine. From Fleming, TR, Issues in the design of clinical trials: Insights from the Herceptin experience (1999). *Seminars in Oncology*, **26** (Suppl. 12): 102–107. Reproduced by permission of Harcourt, Inc.

was 0.017. The study was continued, as these differences were viewed to be of interest but not yet convincing according to statistical monitoring guidelines. At the next review in November 1993, the progression or death endpoint still showed excess events on arm A, and now the survival data had matured to the point that there were 19 deaths on arms A and B. Seventeen of these 19 were on arm A, an eightfold increase in the overall death rate compared to that on arm B; the difference was significant at less than the nominal 0.001 level. In addition, when counting repeated symptomatic AIDS defining events along with the deaths, there were twice as many events (73 vs. 37) on arm A as on arm B.

Since these data strongly suggested that arm A could be therapeutically inferior to arm B, it then was of particular concern to observe that arm A was actually the ddI placebo group, and arm B the ddC placebo group. At the point of this interim analysis, there was only a modest reduction in the rate of progression or death on active ddI (55/337) and on active ddC (62/344) relative to the pooled control group (70/340), and there were equal death rates in the patients on active ddI (18 of 337), on active ddC (18 of 344), and in the overall pooled control group (19 of 340). Thus, while there was evidence for little difference in efficacy of combination chemotherapy versus AZT alone, the two placebo groups differed

Table 5.5 Interim and final results in CPCRA 007 trial. From Fleming, TR, Issues in the design of clinical trials: Insights from the Herceptin experience (1999). *Seminars in Oncology*, **26** (Suppl. 12): 102–107. Reproduced by permission of Harcourt, Inc.

	Regimen		*p*-value
	A	B	
August 1993			
Sample Size	151	151	
Prog./Death	33	16	0.017
Death	8	2	0.11
All Events	54	24	
November 1993			
Sample Size	172	168	
Prog./Death	42	28	0.033
Death	17	2	<0.001
All Events	73	37	
May 1995			
Sample Size	188	187	
Prog./Death	100	95	
Death	75	66	
All Events	210	202	

strikingly. Recognition of this difference prompted a careful examination of the placebo formulations used in the trial. The ddI placebo contained the buffering agent included in the ddI preparation to alter the gastric pH and reduce the inactivation of the drug in the stomach. This led the DMC overseeing the trial to question whether this ingredient in the placebo formulation was truly inactive or perhaps responsible for drug–drug interactions or other unintended effects.

The DMC faced a difficult dilemma. Termination of the ddI placebo group was strongly motivated by the acknowledgment that use of a placebo with plausible potential for meaningful adverse effects could not be tolerated since the placebo would not provide a counterbalancing realistic hope for benefit. On the other hand, termination of the ddI placebo with a suggestion for harm, if not justified, would seriously jeopardize the interpretation and complicate the blinding of the trial, and would jeopardize the interpretation of other major concurrent trials that were employing the same ddI placebo formulation.

Fortunately, an identically designed trial, entitled Delta (Delta Coordinating Committee, 1996), was being conducted concurrently in Europe. The DMCs from CPCRA 007 and Delta agreed to share their key outcome data, agreeing that strict confidentiality of this information would be maintained. Reassured by the lack of differences between the two placebo groups in Delta, the 007 DMC recommended continuation with ongoing monitoring of that study.

Table 5.5 reveals the final results of CPCRA 007 obtained in May 1995. The excess events on the ddI placebo were substantially diminished, although a small non-significant increase in mortality was still observed.

This experience clearly illustrates the potential benefits that can be achieved by sharing of confidential information between two DMCs monitoring concurrent related trials. Even though the number of excess deaths on the ddI placebo became smaller in subsequent analyses, this study also demonstrates the need to ensure that the inclusion of placebos provides neither a substantial inconvenience to the patients in the control arm nor the potential of harm, since there is no expected benefit to be delivered.

5.2.6 Indirect challenges to confidentiality

DMC members involved in patient care and/or medical policy-making may frequently face unusual challenges to maintaining confidentiality of interim results. One colleague describes a situation in which the emerging results from a trial on whose DMC he served suggested that a common treatment for the management of angina was probably suboptimal. At that time, he was the director of a hospital unit that used this treatment routinely. He recognized that any change in practice in his unit would undoubtedly (and correctly) have been interpreted as a statement of the status of the interim study results at that time (Julian, personal communication, 2000).

A similar issue relates to one's personal health care. Emerging results of a trial might suggest to a DMC member that his/her medication regimen might not be optimal. If the potentially preferable regimen were not currently standard practice it might be difficult to discuss a possible change with a personal physician without revealing confidential information – especially if that physician were aware of the patient's role on the DMC of the ongoing trial. This problem could affect non-physician as well as physician members of DMCs.

These sorts of problems highlight the inevitable tensions between patient care and clinical research. Clinical trials are intended to answer specific questions definitively and reliably, but practicing physicians will use all the information they have available – some more reliable than others – to make decisions about treatment of individual patients. These types of dilemmas cannot therefore be avoided, but it is useful for prospective DMC members to understand that they occur.

5.3 THE NEED FOR THE DATA MONITORING COMMITTEE TO REVIEW UNBLINDED DATA

The importance of blinding the trial participants, their care-givers and the study sponsors to interim efficacy and safety data has been justified. Some have argued that blinding should be extended to the DMC membership, thereby further contributing to the protection of study integrity. However, it is scientifically and ethically problematic to withhold from the DMC access to the efficacy and safety data that are fully unblinded by intervention group.

Three arguments have been made in support of providing a DMC with data using treatment codes rather than the actual treatments: first, blinded reports to the DMC would reduce the risk of 'leaks' if this information should fall into the 'wrong hands'; second, the risk of leaks by the DMC would be reduced; and third, limiting the DMC's access through partial unblinding would reduce the risk that this body would overreact to early and potentially misleading results, that is, to something 'not real'.

The first point is the easiest to rebut. The reports sent to the DMC in advance of the meeting, and those presented at the meeting itself, can certainly be printed using treatment codes (e.g., A vs. B) to protect against inadvertent misdelivery or misplacement of reports. The decoding information could be provided to DMC members under separate cover. This approach has the added benefit of permitting DMC members who wish to see the data in coded fashion first, before unblinding themselves, to have the opportunity to do so.

The second point is valid, but weak. Certainly, the fewer individuals with access to the unblinded interim data, the smaller the possibility of a 'leak'. Although it seems unlikely that no DMC member has ever leaked results of an interim analysis, in our combined experience we are not aware of any such leaks by DMC members.

The third point is really the core reason why some believe that the DMC should not be fully unblinded to interim results (Pocock and Furberg, 2001). Here, we strongly disagree. First of all, unless the safety data are coded differently from the efficacy data, blinding cannot usually be accomplished because the type, frequency and severity of adverse events will usually identify the treatments. If the safety data and the efficacy data are coded independently (e.g., using X and Y for treatment codes when efficacy outcomes are reported, and A and B for treatment codes when safety outcomes are reported), the DMC will be unable to make benefit-to-risk assessments, largely undermining its ability to do its job. Second, for adequate protection of trial participants, someone must be aware of the treatment codes. The DMC, having been specifically constituted to be knowledgeable and as free of conflicts of interest as possible, would seem to be the optimal entity to be entrusted with this information. Since its most important responsibility is to protect the interests of study participants, the DMC needs to be fully informed to allow the earliest possible detection of something that is 'real'. Meinert (1998) states:

> Masked monitoring denies the monitors the key information they need to perform in a competent fashion, and incompetent monitoring poses a risk to research subjects.

He goes on to say:

> It is imperative that someone be aware of the nature and trend of the results as randomized treatment trials proceed. If the desire to ensure objectivity keeps the investigators from assuming this role, then it should be assumed by fully informed monitoring committees that perform in accordance with ethical principles and to the satisfaction of institutional review boards.

The initial monitoring of a recent clinical trial having an array of neurological endpoints illustrates these concerns. At the first interim review of outcome data by the trial's DMC, the data analysis center statistician provided reports in which safety and efficacy results for the trial's two treatment groups were coded A/B, with this coding being randomly permuted for each of the myriad of neurological outcome measures. The DMC therefore was unable to assess patterns across outcomes, and was also prevented from making benefit-to-risk evaluations. The DMC insisted it receive unblinded reports. These were needed to achieve timely detection of treatment-related adverse neurological effects that, to be adequately understood, required an integration of complex patterns in the data. The coded reports did not permit assessment of the strength and consistency of evidence across different sources of information, including comparisons between data from the 'serious adverse event' regulatory reporting system and the case report form-based adverse event coding system. The ability to evaluate the quality and completeness of data was also diminished.

The Cardiac Arrhythmia Suppression Trial (CAST) further illustrates concerns arising from blinding the DMC (Echt *et al.*, 1991). The interim analysis results for CAST were presented to the DMC in a blinded fashion, using X/Y coding for the intervention and placebo groups. At the first meeting in which the DMC received interim analyses, a trend already was beginning to emerge, with 13 vs. 7 deaths. Since the DMC was blinded, it was unaware that this trend actually favored placebo. Hence, no arrangements were made by the DMC to alter the previously established plan to wait 6 months for its next review of data. Fortunately, the statistical center did detect that the mortality trend increased rapidly. The DMC was then alerted to the treatment identity through a conference call. A regular meeting was promptly arranged, allowing the DMC to evaluate the entire data set with full knowledge of the treatment identity. As recommended by the DMC at that meeting, the trial was promptly terminated, but not before the excess mortality had become 56 vs. 22. It is not clear that anything useful was gained by keeping the DMC for the CAST trial blinded at its first review of interim analysis results. Meanwhile, it is apparent that this blinding limited the time the DMC had to provide a thoughtful response to this rapidly emerging trend, potentially delayed this response, and resulted in placing considerable responsibility solely on the judgment of the statistical center statistician.

While maintaining the blinding of interim efficacy and safety data is critical to trial integrity, it is improper to blind the DMC itself. The primary responsibility of the DMC is to safeguard the interests of study participants. Meeting this responsibility leads to an ethical imperative that the DMC have timely access to unblinded data on all relevant treatment outcomes, to enable the earliest possible detection of evidence that establishes a study regimen to have an inferior benefit-to-risk profile.

REFERENCES

Abrams D, Goldman A, Launer C *et al.* (1994) A comparative trial of didanosine or zalcitabine after treatment with zidovudine in patients with human immunodeficiency virus infection. *New England Journal of Medicine* **330**: 657–662.

Armitage P on behalf of the Delta Data and Safety Monitoring Committee (1999) Data and safety monitoring in the Delta Trial. *Controlled Clinical Trials* **20**: 229–241.

Brosgart CL, Louis TA, Hillman DW *et al.* (1998) A randomized, placebo-controlled trial of the safety and efficacy of oral ganciclovir for prophylaxis of cytomegalovirus disease in HIV-infected individuals. Terry Beirn Community Programs for Clinical Research on AIDS. *AIDS* **12**(3): 269–277.

Delta Coordinating Committee (1996) Delta: A randomized double-blind controlled trial comparing combinations of zidovudine plus didanosine or zalcitabine with zidovudine alone in HIV-infected individuals. *Lancet* **348**: 283–291.

DeMets DL, Fleming TR, Whitley RJ *et al.* (1995) The data and safety monitoring board and acquired immune deficiency syndrome (AIDS) clinical trials. *Controlled Clinical Trials* **16**: 408–421.

Dixon DO, Lagakos SW (2000) Should data and safety monitoring boards share confidential interim data? *Controlled Clinical Trials* **21**(1): 1–6.

Echt DS, Liebson PR, Mitchell LB *et al.* (1991) Mortality and morbidity in patients receiving encainide, flecainide, or placebo. The Cardiac Arrhythmia Suppression Trial. *New England Journal of Medicine* **324**: 781–788.

Fleming TR, DeMets DL (1993) Monitoring of clinical trials: Issues and recommendations. *Controlled Clinical Trials* **14**: 183–197.

Fleming TR, Ellenberg SS, DeMets DL Monitoring clinical trials: issues and controversies regarding confidentiality. *Statistics in Medicine*, in press.

Green SJ, Fleming TR, O'Fallon JR (1987) Policies for study monitoring and interim reporting of results. *Journal of Clinical Oncology* **5**: 1477–1484.

Marx JL (1989) Drug availability is an issue for cancer patients, too. *Science* **245**: 346–347.

Medical Research Council Working Party (1984) The evaluation of low-dose preoperative x-ray therapy in the management of operable rectal cancer: results of a randomly controlled trial. *British Journal of Surgery* **71**: 21–25.

Meinert CL (1998) Masked monitoring in clinical trials – blind stupidity? *New England Journal of Medicine* **338**(19): 1381–1382.

Moertel CG, Fleming TR, Macdonald JS, Haller DG, Laurie JA, Goodman PJ, Ungerleider JS, Emerson WA, Tormey DC, Glick JH, Veeder MH, Mailliard JA (1990) Levamisole and fluorouracil for adjuvant therapy of resected colon carcinoma. *New England Journal of Medicine* **322**: 352–358.

Pocock S, Furberg CD (2001) Procedures of data and safety monitoring committees. *American Heart Journal* **141**: 289–294.

Rider WD, Palmer JA, Mahoney LJ, Robertson CT (1977) Preoperative irradiation in operable cancer of the rectum: Report of the Toronto trial. *Canadian Journal of Surgery* **20**: 335–338.

Saravolatz LD, Winslow DL, Collins G *et al.* (1996) Zidovudine alone or in combination with didanosine or zalcitabine in HIV-infected patients with the acquired immunodeficiency syndrome or fewer than 200 CD4 cells per cubic millimeter. *New England Journal of Medicine* **335**: 1099–1106.

Spector SA, McKinley GF, Lalezari JP *et al.* (1996) Oral ganciclovir for the prevention of cytomegalovirus disease in persons with AIDS. *New England Journal of Medicine* **334**(23): 1491–1497.

6

Data Monitoring Committee Meetings

Key Points

- Standard operating procedures for a DMC should be established.
- Early in the trial, DMC review will focus more on safety, quality of conduct and trial integrity than on efficacy evaluation.
- Interim data reports submitted to the DMC should be as accurate and as up to date as it is feasible to accomplish.
- DMC meetings may include open and closed sessions, with trial leadership permitted to attend open sessions and comparative interim data presented and discussed only in closed sessions.
- Minutes should be kept for all DMC sessions.

6.1 INTRODUCTION

The DMC activities required to safeguard the interests of trial participants and to preserve trial integrity are primarily conducted during its meetings. These activities include reviewing the scientific design of the trial and proposed operating procedures, monitoring early data regarding safety and quality of trial conduct, and providing in-depth reviews of efficacy and adverse effects on an ongoing basis as the study matures. The meetings not only allow the DMC to discuss with each other any issues or concerns arising from their review of the study reports that are provided in advance, but can also allow the DMC to interact with study investigators and sponsor(s).

As noted in Chapter 2, a DMC charter describing standard operating procedures (SOPs) should be established in the planning stages of a clinical trial in order to ensure that the DMC meetings are conducted in an efficient and effective manner.

Some of these SOPs relate to defining:

- the specific objectives and the timing of the various types of meetings of the DMC;
- the procedures to be followed in preparation of reports to be presented to the DMC;
- the organization and format for the DMC meetings, including the procedures for any planned interaction with study investigators and sponsors; and
- the development and circulation of meeting minutes.

This chapter will address important elements of these SOPs.

6.2 SPECIFIC OBJECTIVES AND TIMING OF MEETINGS

The specific objectives of the DMC meetings evolve during the course of the planning and conduct of the trial. While the primary function of the DMC is carried out during the trial conduct stage, the committee may also have responsibilities that begin before initiation of recruitment of study participants, and that could extend beyond trial termination.

6.2.1 Organizational meeting

The first meeting of the DMC should take place before the study begins if possible. It should be an organizational meeting, at which the protocol can be discussed, plans and procedures can be established, and the members can become acquainted with each other and with key members of the study organization. Typical agenda items at the organizational meeting of the DMC would include:

1. introducing the DMC members to each other and to lead representatives from the steering committee and/or investigators and the study sponsor;
2. discussing the penultimate version of the study protocol and formulating any comments or recommendations relating to ethical, scientific or practical concerns;
3. discussing and making any necessary modifications to the SOPs for the role and functioning of the DMC; and
4. developing recommendations for the format and content of the reports that will be used to present trial results at future DMC meetings.

Item 1 is self-explanatory. The rationale for item 2 was discussed in Chapter 2. Briefly, DMC members must be ethically and scientifically supportive of the trial's design if they are to be able to carry out their primary responsibilities of safeguarding the interests of trial participants and preserving the integrity and credibility of the trial. Thus, they should have the opportunity to review the study protocol at the time of completion of its penultimate draft.

Regarding agenda items 3 and 4, it is advisable for the trial leadership, together with the DMC chair and DMC statistician, to prepare initial drafts of the SOPs regarding the role and functioning of the DMC, often referred to as the DMC Charter, and of the format and content of the open and closed reports. Appendix A provides a sample draft of the DMC Charter, while Tables 5.1 and 5.2 in the previous chapter provide typical outlines for the content of the open and closed reports that would be appropriate for most trials.

The DMC Charter, as discussed in Chapter 2, should define the primary responsibilities of the DMC, its composition, the timing and purpose of its meetings, the nature of its interactions with the trial's sponsor and/or steering committee (when there is no separate steering committee charter), the statistical monitoring guidelines to be implemented, and an outline for the content of the DMC's open and closed reports (see Chapter 5).

As discussed in section 5.2.1, the open report would include data, pooled by treatment regimen, on baseline characteristics as well as on aspects relating to quality of study conduct, such as recruitment progress, eligibility violations, trial adherence, and currency and completeness of follow-up (see Table 5.1). The open report is to be made available to all who attend the open session of the DMC meeting, as discussed in section 6.4.2 below. In contrast, the closed reports would provide confidential information to the DMC, including a display by intervention group for all data elements that had been presented only in the aggregate in the open report. The closed report would also include analyses of the primary and secondary efficacy endpoints, subgroup and adjusted analyses, detailed analyses of adverse events, symptom severity and other relevant safety data, and analyses of lab results (see Table 5.2).

In preparation for this organizational meeting, it can be useful to generate sample pages for each of the tables and figures that have been proposed to appear in the open and closed reports. By viewing artificial data in these tables and figures, the DMC will be able to more effectively identify and recommend refinements in the content and format of these reports in advance of the DMC meetings at which they will perform safety reviews and reviews of formal interim efficacy analyses. Although it is likely that some modifications and additional analyses will be needed during the course of the trial no matter how much effort goes in to the initial planning, careful planning upfront can minimize the need for changes and improve both the efficiency and effectiveness of the DMC's work.

6.2.2 Early safety/trial integrity reviews

Once recruitment and protocol-specified interventions have been initiated, monitoring of safety outcomes must be conducted on a regular basis. While serious adverse events are reported immediately to regulatory authorities by the trial's medical monitor, such reporting usually does not include information on the assigned treatment; only the DMC has access to accumulating adverse event data aggregated by intervention group.

In settings where there is considerable risk for rapidly emerging adverse events, the chair or a designated clinical member of the DMC may be provided serious adverse event data on a relatively frequent basis – monthly or even weekly. Independent assessment of these individual case reports may sometimes be an important supplement to the work of the sponsor's medical monitor.

The entire DMC will be engaged in the regularly scheduled safety review in order to adequately ensure early detection of unacceptable safety risks to study participants. Since safety risks often become apparent well before a substantial fraction of efficacy endpoints occur, one or more meetings of the DMC to conduct early safety reviews are generally held before the time of the first formal interim efficacy analysis. If important safety risks are detected, the DMC could recommend modifications to the conduct of the trial, such as adjustments to the dose or schedule of the interventions or to the frequency and intensity of monitoring, or modifications to the risk groups eligible for the trial. Notifications could be provided to institutional review boards, investigators, regulatory authorities, care-givers or the study participants. Recommendation might also be made for temporary cessation of recruitment or, in the most serious instances, for termination of the trial.

When safety risks are detected, it often is a difficult challenge for a DMC to arrive at its recommendation for a proper course of action. This is especially true in settings addressing the treatment or prevention of diseases that induce risks for major morbidities or death, since some increase in safety risks could be acceptable if there is evidence or likelihood that greater efficacy would be achieved by the experimental intervention. It follows that the DMC should be provided available efficacy as well as safety data at any of these early safety review meetings where benefit-to-risk assessments would be important considerations in formulating recommendations to address emerging safety risks.

The motivation for providing the DMC access to efficacy data at early safety reviews – a motivation often not fully appreciated by trial sponsors or organizers – is typically not to consider early termination based on compelling evidence of benefit, but rather to enable benefit-to-risk assessments in the presence of early emerging safety concerns. Nevertheless, regulatory authorities and others could require adjustments to be made to account for potential increases in the false positive error rate induced by these additional 'looks' at the efficacy data. Fortunately, this issue is readily addressed by having a standard group sequential monitoring boundary in place that is very conservative when only a small fraction of efficacy endpoints are available. The alpha adjustments required are truly negligible, even when one has conducted several early safety/trial integrity reviews involving access to both efficacy and safety data.

DMC consideration of early administrative data is also advisable (Fleming, 1993). Factors such as accrual rates, compliance with eligibility restrictions, covariate balance by intervention group, adherence to protocol specifications for delivery of the interventions, control arm event rates for the primary efficacy endpoint, quality of data capture, and completeness of follow-up on outcome measures may impact on the trial's ability to provide a reliable answer to the question of interest. Thus, evaluation of such factors cannot be delayed until the time of formal interim efficacy analyses in long-term studies. The early safety/trial integrity reviews provide the DMC the opportunity to identify concerns relating to trial conduct as soon as possible, and to provide recommendations that will enhance the quality of the trial and reliability of study conclusions.

During the early safety/trial integrity review meetings, the DMC may see the need for additional refinements in the content and format for the open and closed reports to be generated at the time of the DMC's review of the formal interim efficacy analyses. This exercise for developing refinements may sometimes be conducted, at least in part, in the open session, in order for the trial leadership to have input on the types of analyses to be performed. In such cases, one must be careful that the figures and tables are generated in a manner that maintains appropriate confidentiality of available information. Specifically, artificial data could be used to create the proposed tables and figures that will present the efficacy and safety data. For tables and figures presenting information that would appear by intervention group in the closed report but only in the aggregate in the open report (such as data on recruitment, adherence, retention and quality of data capture) actual data can be used except that each participant should be assigned an artificial randomly selected intervention group assignment.

6.2.3 Formal interim efficacy analyses

At the meetings of the DMC in which formal interim efficacy analyses are reviewed and evaluated, the committee should receive comprehensive information from the open and closed reports regarding the relative safety and efficacy of the regimens being assessed. The currentness, completeness and accuracy of information provided by the reports should meet the standards to be defined in section 6.3. At these meetings, as will be described in section 6.4, the DMC may also obtain useful insights from interactions with the members of the trial leadership who are in attendance at the open sessions of the meeting.

As it reviews the formal interim efficacy and safety analyses, the DMC will weigh evidence for benefits and risks to determine whether there remains adequate balance of benefit to risk in order to ethically and scientifically justify trial continuation. In most settings, it will be proper and advisable to continue the trial. However, even in such settings, modifications to the study regimens or trial procedures might appear advisable to enhance the safety of trial participants. Recommendations for modifications could extend to the oversight procedures themselves. While the timing of formal interim efficacy analyses is usually identified in the study protocol, the DMC might recommend changes to the timing of future formal analyses of efficacy and safety data, on the basis of data patterns observed at early analyses.

At the completion of each formal interim efficacy analysis, the DMC may provide recommendations regarding:

- continuation or termination of the trial;
- modifications to study regimens and/or oversight procedures to enhance participant safety;
- modifications to procedures for participant recruitment or management or to data capture procedures to enhance trial quality and integrity; and
- modifications to informed consent or other information provided to participants.

6.2.4 End-of-trial debriefing

As discussed in Chapter 2, the trial leadership will conduct its review of unblinded data once follow-up and final changes to the database have been completed. In this process, the sponsor and/or study chair may invite the DMC to contribute its unique insights obtained through its unblinded review of efficacy and safety data throughout the conduct of the trial. It is common to acknowledge the DMC in manuscripts describing the main study results. It is generally inappropriate, however, for members of the DMC to be co-authors of the manuscript that provides the primary results of the trial. In some cases, the monitoring process itself may be of sufficient methodological interest to be described in a paper. Such papers serve a useful purpose. They may appropriately include the DMC as authors but should generally be subject to the same scrutiny by the study leadership as any other papers arising from the study.

6.3 PREPARATION OF MEETING REPORTS

At all meetings at which data are reviewed, it is imperative that information on safety and efficacy provided to the DMC be as accurate, complete and timely as possible. To the extent they are otherwise, the ability of the DMC to make informed judgments about the appropriateness of trial continuation or modifications to trial conduct will be compromised.

Example 6.1: ACTG 019: Zidovudine (AZT) monotherapy in asymptomatic HIV-infected patients

The AIDS Clinical Trials Group (ACTG) initiated trial 019 in August 1987 to evaluate whether AZT would slow the progression of disease in asymptomatic HIV-infected individuals (Volberding *et al.*, 1990). In August 1989, the DMC

Table 6.1 ACTG 019 clinical trial: analysis of August 2, 1989 (data freeze on May 10, 1989) and updated analysis of August 16, 1992. From Fleming, 1992

	Treatment arm	No. of Progressions	Progression* rate	p-value vs. placebo
Aug. 2, 1989	Placebo (428)	31	7.5	
	500 mg (453)	8	2.1	0.0008
	1500 mg (457)	12	3.4	0.015
Aug. 16, 1989	Placebo (428)	38	7.6	
	500 mg (453)	17	3.6	0.0030
	1500 mg (457)	19	4.2	0.05

* Progressions per 100 person-years of follow-up

reviewed data on 1338 patients who had been followed for an average of 12 months post randomization.

The history of interim results for these patients is summarized in Table 6.1. The data presented at the August 2, 1989 DMC meeting were current only through May 10, 1989, 3 months earlier. Interim results suggested a favorable trend for both low- and mid-dose AZT arms relative to placebo. These results for the low-dose regimen satisfied the O'Brien–Fleming group sequential guideline for early termination of the placebo. In addition, at the open session of the DMC meeting, the study investigators reported serious reservations about the value of trial continuation to assess longer-term effects since an increasing percentage of placebo patients were choosing to initiate active therapy, even in the absence of definitive data.

Two issues concerned the DMC. First, the data were not sufficiently current; the events occurring between May and August could substantially alter the impressions about treatment effect. Second, the data may not have been fully accurate, as the study team had not yet verified all reported outcome events. Thus, the DMC delayed making a recommendation regarding trial continuation or termination until these two issues could be clarified. The protocol team was asked to update the report for primary outcome measures (through August 1) and to verify that all events fulfilled the protocol definition.

This task was completed within 2 weeks and a conference call for all DMC members was held shortly thereafter. Updated follow-up identified nearly 50% more patients having had documented symptomatic AIDS-defining events. The differences in the rate of outcome events between the low-dose and placebo groups still met the monitoring boundary criterion for termination, (Table 6.1, $p < 0.005$).

The updated information did alter the interpretation of the treatment effect, as estimated by Kaplan–Meier time-to-event curves (see Figure 6.1). For the placebo versus low-dose ADT comparison, the curves on August 2 (Figure 6.1a) suggest a sustained reduction in the hazard rate over time. There is a suggestion that AZT would provide a very substantial improvement in long-term survival. In contrast, the curves presented on August 16, in Figure 6.1b, suggest that the AZT curve is essentially a 6-month translation of the placebo curve. Thus, while the DMC did recommend termination of the placebo arm of the trial, it also reported to investigators that the data were more consistent with a delay in disease progression than evidence of a cure.

The ACTG 019 experience motivates implementation of procedures to maximize the currency, accuracy and completeness of data presented to the DMC. The following outline identifies important elements of such procedures. The suggested timeline would apply to trials having duration from initiation of participant accrual to study closure that would be between 12–18 months and 3–4 years.

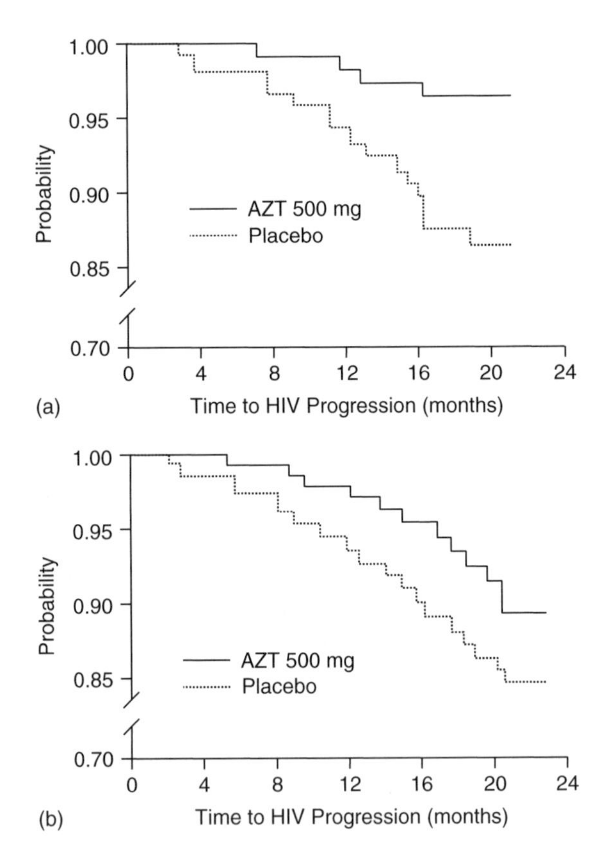

Figure 6.1 Kaplan–Meier estimates of the probability of being free of progression to AIDS-defining events or death, as presented at the DMC meeting on (a) August 2, 1989, (b) August 16, 1989. From Fleming, 1992.

More stringent and more lenient timelines might be appropriate for trials of very short and very long duration, respectively.

First, establish the date of the DMC meeting (referred to as 'day 0' in this timeline) and then compute the 'date of data freeze', which should be approximately day −60. The goal will be to have data and analyses in the meeting reports that will be as complete and accurate as possible regarding inclusion of patient visits and trial safety and efficacy outcomes that had occurred by the date of data freeze, to the extent possible.

Second, ensure that data collection procedures are designed to optimize the currency of data capture as one approaches the date of data freeze. Whatever the mode of data submission – electronic, telefax, mail or capture by contracted clinical research associates who visit study sites – the timing of submissions should be coordinated with the timing of the DMC meetings. The sites and CRAs should plan their submissions and site visits to maximize the completeness of the data that would be sent to the statistical center by the date of data freeze.

Third, between the date of data freeze and day −25, the iteration of queries and responses between the data management center and study sites should be completed, ensuring that the database has the most accurate and complete data through the date of data freeze that it was possible to obtain.

During these five weeks after the date of data freeze, the sites, CRAs and the data management center should work together toward this goal of achieving complete and accurate capture data relating to patient visits and trial outcomes that had occurred through the date of data freeze.

Fourth, at day −25, the statistical database corresponding to data through the date of data freeze should be locked and statistical analysis files should be generated and provided to the statisticians at the statistical center responsible for generating analyses and the open and closed reports.

As noted earlier, the format and content of the open and closed reports for the DMC should be established at the organizational meeting of the DMC and should be refined at the time of the early safety/trial integrity reviews. By the date of data freeze, the statistical center should have finalized the statistical software required to generate the open and closed reports for the DMC, and should have tested this software using artificial data.

Fifth, between day −25 and day −7, the statistical center should conduct the appropriate analyses and generate the open and closed reports.

Finally, the open and closed reports should be sent to the DMC in a manner to ensure they are received at least 3–4 days in advance of the DMC meeting. In some settings, DMC members might require more time to review the DMC reports. But there is always an important balance to consider. While it is of great importance to ensure that the DMC members have ample opportunity to become thoroughly familiar with the content of these reports, it must be kept in mind that currency and completeness probably would be compromised if the committee members were to receive the reports a week or more before the date of the DMC meeting. DMC members can effectively contribute to this currency as well as to the quality of their oversight by setting aside ample time to review these reports during the last few days before the DMC meeting.

Even if these procedures and this timeline are met for ensuring accurate and complete data through the date of data freeze, it is still likely in many trials that important information will have evolved on the trial's primary endpoint(s) between the date of data freeze and the date of the DMC meeting. For example, some endpoints may have occurred prior to the date of data freeze but were not adjudicated until after the date the database was locked; other endpoints may have occurred after the date of data freeze and may have been subsequently reported. Thus, in many trials it has been standard practice for the statistical center to take a 'snapshot' of the primary endpoints in the database approximately 7–14 days before the date of the DMC meeting. Some investigators urge that all available primary endpoints should be included in the interim analyses even though not all are fully adjudicated (Pocock and Furberg, 2001; Wittes, 2000). Even when the primary analysis focuses on adjudicated endpoints, a separate brief summary could be generated that would provide a more current (though less cleaned or validated) update of the primary endpoint analysis. (Had this been done, for example, in the ACTG 019 trial discussed earlier in this section, it might have been possible for that DMC to reach a decision about trial termination at its in-person meeting on August 2, 1989.)

6.4 FORMAT FOR MEETINGS

The meetings of the DMC should be conducted in a manner to ensure that committee members:

- fully understand the information provided in the open and closed reports;
- can benefit from the insights provided by representatives of the investigators and the study sponsor;
- can identify issues that remain inadequately addressed and develop a strategy to obtain necessary additional information; and
- can develop a consensus regarding the appropriateness of continuing the trial and the recommendations to be made to the trial leadership to enhance the quality of the trial design and conduct.

A variety of approaches to structuring the meeting are possible. One format that achieves the objectives above, while enabling the DMC to preserve confidentiality of study results, involves the conduct of several successive sessions. In this format, study representatives would engage in discussions with the DMC in an open session, and the DMC would consider unblinded comparative data in closed sessions that have attendance limited to the DMC membership (DeMets *et al.*, 1995). We have found this format to be very effective, and describe it in greater detail.

6.4.1 The closed session

The DMC members, in advance of their meeting, should have carefully reviewed the open and closed reports. These will provide insights regarding recruitment, eligibility violations and baseline characteristics, and data on adherence to treatment and completeness of follow-up (presented in the aggregate in the open report and by intervention group in the closed report), and closed report information regarding unblinded comparative analyses of primary and secondary endpoints, analyses of adverse events and symptom severity, and laboratory results. The closed session then provides the DMC the opportunity to discuss this information, to develop initial perspectives about issues regarding trial conduct and safety and efficacy of interventions, and to identify issues to be addressed during the remainder of the meeting.

This closed session usually should be attended only by members of the DMC and by the statistician responsible for conducting the analyses and generating the open and closed reports. This presenting statistician provides an essential linkage between the committee and the database, and is able to answer questions relating to interpretation of the reports. (The advantages of this statistician being independent of the sponsor and trial leadership are discussed in Chapter 7.)

In some settings (particularly when a DMC is reviewing multiple trials at its meeting), the DMC chair may appoint a primary clinical and a primary statistical reviewer of the open and closed reports (or of specific sections of the reports) at the time the reports are circulated to the committee. These individuals can then efficiently lead the committee through a discussion of the most important findings. After the full committee membership has had an opportunity for careful discussion of issues of importance, concern or uncertainty, the DMC can formulate a list of issues to be discussed with the members of the sponsor and those investigator representatives who will attend the open session. The DMC can also formulate a list of any additional analyses they would like to have generated by the presenting statistician at the next meeting (or earlier).

6.4.2 The open session

The open session provides a forum for exchange of information among the various parties who share responsibility for the successful conduct of the trial. Thus, the DMC members and the presenting statistician are joined by the lead trial investigator(s) and representatives of the sponsor, and potentially by regulatory authorities.

The protocol chair or other investigators may be asked to provide a brief summary of the study progress, including recruitment, quality of data, and other issues raised in the open report, that usually would have been provided to the sponsor and investigator representatives by the time of the DMC meeting. The protocol chair should be given the opportunity to discuss problem areas of which the DMC should be aware and would have the opportunity to seek advice or support from the Committee regarding any planned actions to solve these problems. The sponsor and regulatory representatives would also have the opportunity to comment or query those in attendance and might wish to bring specific issues to the attention of the DMC. In some cases, for example, the sponsor may wish to share with the DMC data from one or more ongoing studies that are relevant to the current trial.

The open session also provides the DMC the opportunity to query the sponsor and lead investigators regarding issues identified during the initial closed session. Such queries often relate to possible modifications in trial design or conduct that the DMC might wish to suggest in their final recommendations. Examples of such issues include the need

- to improve adherence to regimens;
- to avoid excessive accrual by a single institution;
- to improve the quality or timeliness of data capture;
- to determine why missingness of certain data elements or loss to follow-up is high; or
- to reduce safety risks through modified dosing schedules or dosage adjustment algorithms or through imposing additional eligibility restrictions to avoid exposing participants who have baseline characteristics found to be associated with high safety risks.

The open session should be conducted in a manner that fully maintains confidentiality of all information provided in the closed report, including results from analyses of efficacy and safety data. DMC members should take care to avoid conveying information by asking questions, making comments, or even exhibiting 'body language' that could suggest the emergence of certain patterns in the interim data.

At times, the most important insights the DMC might wish to obtain during the open session would be the perspectives the sponsor and investigators would have about the relative benefit-to-risk profile of the trial regimen had they been unblinded to the current information in the closed report. Obtaining these perspectives without unblinding these individuals is difficult, but helpful information can sometimes be obtained through well-formulated questions.

An excellent example is provided by the ACTG 981 trial (Powderly *et al.*, 1995), discussed in Chapter 2, which evaluated fluconazole for the prevention of serious fungal infections in AIDS patients. While a highly significant benefit was found on the primary endpoint at an interim analysis, the fluconazole-treated patients surprisingly experienced a substantial increase in the rate of death. At the open session, the DMC provided the investigators a series of scenarios relating to the effects of fluconazole on death, serious fungal infections and other endpoints, and asked them to comment on how they would view the benefit-to-risk profile in each setting. Although the investigators were surely aware that some decision process was under way, the DMC felt it was able to obtain the investigators' perspectives without unblinding them to the current data.

6.4.3 The final closed session

The final closed session is attended by the DMC members and the presenting statistician. (At times, an executive closed session, attended only by members of the DMC, is preferred, or may be added.) At this final closed session, with insights obtained from the previous sessions, the DMC should develop a consensus on its list of recommendations, including that relating to whether the trial should continue. It is far preferable, and usually achievable, for the DMC to seek and arrive at consensus on each of its important recommendations rather than to settle for a simple majority vote of approval.

6.4.4 Various formats for holding the open and closed sessions

Various formats have been employed for ordering the open and closed sessions of the DMC meeting. The closed session/open session/final closed session format just described allows the DMC to develop its initial perspectives on the data independent of input from the sponsor and/or investigators during the open session, and to ensure that adequate time will be allocated to closed session discussions. This format also allows the DMC to be better prepared for the open session, with a consensus of the most important issues to be addressed with the sponsor and investigators in attendance at that session. This format would be especially appropriate when only a half-day or less would be available for the entire DMC meeting. It also enables efficient engagement of the study investigators and sponsor representatives when they are available for only a limited block of time.

A somewhat more complicated version of this format has been used in settings in which a somewhat longer interval is available for the meeting and the sponsor and investigators are available throughout that time. In this variation, additional brief open sessions are held at the beginning and end of the meeting. The initial brief open session allows the sponsor and investigators to provide some initial perspectives about the status of the trial and possibly to provide a list of questions to the DMC. At the final open session, the sponsor and investigators receive the DMC recommendations in person.

A more common format is one that includes only two sessions: the DMC may meet with sponsors and others initially in an open session, and then retreat to a closed session to review the comparative data. This format is simple, and frequently implemented. While it is generally workable, this approach does not allow the DMC to have initial time on its own to discuss the issues raised by the interim reports and to develop questions it might wish to address to the sponsor and any other attendees at the meeting's primary open session.

6.4.5 Meeting duration and venue

The amount of time that should be set aside for a DMC meeting may vary substantially, depending on the purpose of the meeting and the status of the study at that time. For example, the organizational meeting, at which the DMC members and the trial leadership meet each other, discuss the trial protocol and administrative procedures, the DMC charter, proposed templates for interim reports and other related issues, may require more time than some later meetings at which initial safety data and aspects of trial progress and quality of conduct are considered but comparative efficacy data are not yet mature enough to warrant formal review. There will always be a tension between setting aside sufficient time for full and thoughtful discussion of the issues before the committee, and limiting the time demands on DMC members and others involved in the meeting. The difficulty of scheduling meetings around the calendars of inevitably very busy people also works against long (i.e., multiday) meetings. We have found that when necessary, satisfactory meetings can be conducted in as short a time as 2–3 hours, but that a minimum of 4–6 hours may be required to adequately address complex emerging issues, particularly when major comparative efficacy analyses are presented for consideration.

In-person meetings of the DMC usually are strongly preferred since they allow more effective interaction and consensus development. However, for some DMC meetings held for early safety/trial integrity review, when the entire meeting may take only 2 hours and no controversial issues are expected to arise, a teleconference may allow more efficient use of time and effort. Additionally, when rapidly emerging trends or new safety concerns require an unplanned DMC meeting, it may not be feasible to arrange an in-person meeting at very short notice. Meeting by teleconference in such situations may be necessary. Some administrative issues may even be amenable to handling by correspondence; for example, consideration of proposed templates for additional interim reports, or assessment of additional analyses requested at a meeting that show expected results (i.e., analyses requested 'just to be absolutely sure' that no problem is emerging).

6.5 MINUTES

The proceedings of the DMC meeting, including the data considered, the deliberations, and the recommendations by the committee, should be recorded in carefully developed minutes. Two sets should be prepared: the open minutes and the closed minutes.

6.5.1 The open minutes and the closed minutes

The open minutes should describe the proceedings in the open session(s) of the DMC meeting, and should summarize all recommendations by the committee.

Since these minutes will be circulated immediately to the sponsor and to lead study investigators, it is necessary that the minutes do not unblind the efficacy and relative safety data if the DMC is not recommending early termination.

The closed minutes should describe the proceedings from all sessions of the DMC meeting, including the listing of recommendations by the committee. Because it is likely that these minutes will contain unblinded information, it is important that they are not made available to anyone outside the DMC. Rather, copies should be archived by the DMC chair and by the statistician preparing the interim reports, for distribution to the sponsor, lead investigators and regulatory authorities at the time of study closure.

The sponsors should routinely provide a complete collection of open and closed minutes to regulatory authorities at the time of new drug applications and biologic licensing applications.

6.5.2 The Level of Detail

There certainly is flexibility in the level of detail for these minutes. It is unnecessary to provide a verbatim transcript of the meeting. On the other hand, there should be adequate detail to provide a clear understanding of the major issues discussed, of the important new information presented that did not appear in the open and closed reports, and of the rationale for the DMC's recommendations.

Since the open report is already available to the sponsor and investigator at the time of the DMC meeting, and since the closed report will be archived and distributed along with the closed minutes at the time of trial closure, it is not necessary for the minutes to provide a detailed repetition of the data in those reports.

For a DMC meeting of 5–6 hours at which a formal interim analysis is reviewed, the closed minutes might typically be three to five pages in length. For each closed session, the minutes should provide the DMC's evaluation of the quality of trial conduct and its impressions about the relative efficacy and relative safety of the study interventions. The additional information that the DMC wishes to obtain should documented, as well as the listing of issues the DMC wishes to pursue with the sponsor and study investigators in the open session. The rationale for any recommendations should be clearly explained.

For the open session(s), the minutes should summarize important new information provided to the DMC by the sponsor and study investigators, as well as their queries to the DMC. The minutes should then summarize the DMC's response to these queries, as well as the discussion held in the open session addressing the list of issues that the DMC had targeted during its initial closed session for discussion with the trial leadership. The minutes of the open session should include all recommendations made by the DMC regarding the conduct of the study.

6.5.3 The authorship of the minutes, and the sign-off by committee members

In many settings, the presenting statistician or a member of that statistician's staff will develop the first draft of the minutes. Ideally, this would be the independent statistician as described in Chapter 7. At times, the DMC chair or another DMC member will assume the responsibility, although this 'double duty' can be rather challenging and may prove unacceptably distracting. For government-funded trials it has been common for the government program representative to draft the minutes, but this also creates other problems, as described in Chapter 4. After the first draft is completed, it is typically circulated to the DMC chair and DMC statistician for their revisions. The draft then is circulated to the entire DMC for their review. When a final version is completed, it should be sent to all DMC members for their sign-off.

The completed version of the open minutes should be sent immediately to the trial leadership to facilitate a timely response to DMC recommendations. The final version of the closed minutes should be archived.

This process for obtaining a completed version of the minutes usually would require at least 1–2 weeks. As a result, as noted in section 6.4.4 of this chapter, a brief final open session is often held to provide the trial leadership an immediate summary of the DMC recommendations. In this final session, the DMC would usually be joined by the study sponsor and lead investigators.

REFERENCES

DeMets DL, Fleming TR, Whitley RJ, Childress JF, Ellenberg SS, Foulkes M, Mayer KH, O'Fallon J, Pollard RB, Rahal JJ, Sande M, Straus S, Walters L, Whitley-Williams P (1995) The data and safety monitoring board and acquired immune deficiency syndrome (AIDS) clinical trials. *Controlled Clinical Trials* **16**: 408–421.

Fleming TR (1992) Evaluating therapeutic interventions: Some issues and experiences (with discussions and rejoinder). *Statistical Science* **7**(4): 428–456.

Fleming TR (1993) Data monitoring committees and capturing relevant information of high quality. *Statistics in Medicine* **12**: 565–570.

Pocock S, Furberg CD (2001) Procedures of data and safety monitoring committees. *American Heart Journal* **141**: 289–294.

Powderly WG, Finkelstein DM, Feinberg J, Frame P, He W, van der Horst C, Koletar SL, Eyster ME, Carey J, Waskin H, Hooton TM, Hyslop N, Spector SA, Bozzette SA (1995) A randomized trial comparing fluconazole with clotrimazole troches for the prevention of fungal infections in patients with advanced human immunodeficiency virus infection. *New England Journal of Medicine* **332**: 700–705.

Volberding PA, Lagakos SW, Koch MA, Pettinelli C, Myers MW, Booth DK, Balfour HH, Reichman RC, Bartlett JA, Hirsch MS, Murphy RL, Hardy D, Soeiro R, Fischl MA, Bartlett JG, Merigan TC, Hyslop NE, Richman DD, Valentine FT, Corey L and the AIDS Clinical Trials Group of the National Institute of Allergy and Infectious Diseases (1990) Zidovudine in asymptomatic human immunodeficiency virus infection. *New England Journal of Medicine* **322**: 941–949.

Wittes J (2000) Data safety monitoring boards: a brief introduction. *Biopharmaceutical Report* **8**: 1–7.

7

Data Monitoring Committee Interactions with Other Trial Components or Related Groups

Key Points

- DMCs will interact, occasionally or routinely, with other trial components, such as the trial sponsor and the study chair.
- Different models for such interaction have been developed, but problems can arise when these other components are provided access to comparative interim data.
- An independent statistician, separate from the primary (steering committee) statistician for the trial, can be a useful addition to the trial structure by preserving the ability of the trial statistician to participate in unbiased interim decision-making with other members of the trial leadership.
- Sharing of interim data between DMCs monitoring similar trials can be valuable but needs to be done judiciously.

7.1 INTRODUCTION

A data monitoring committee needs to interact with many other organizational components of the clinical trial it is monitoring, on a regular or occasional basis. There may be circumstances in which it may wish to, or be asked to, interact with other groups as well. It is useful to consider the nature and extent of these interactions, and to examine how and why these may differ from trial to trial.

7.2 STUDY SPONSORS

The sponsor of the study, usually a pharmaceutical company or a government agency, has ultimate responsibility for the way the study is carried out. Thus it usually appoints DMC members. ('Sponsor' generally connotes the organization or individual with both regulatory and financial responsibility for the study. When these responsibilities are shared, the one with primary regulatory responsibility – that is, the entity holding the investigational new drug (IND) or investigational device exemption application – would usually be the one to direct the study and appoint the DMC. When the IND holder is a government agency, these authorities are often delegated to an external leadership group such as a steering committee.) Since the sponsor must ensure procedures are in place to safeguard the interests of study participants, and is also investing the financial resources the study requires, it will need to have confidence in the group of individuals who will take sole responsibility for reviewing the accumulating interim data. Suppose an inappropriate action is taken by the DMC, such as recommending continuation of a study regimen even though existing data adequately establish it to be inferior, or recommending early termination for efficacy on the basis of data not yet strong enough to be persuasive to the medical community, regulatory agencies, etc. The sponsor will be ethically responsible that trial participants were at undue risk in the former setting, and will have to bear the financial loss of an inconclusive study in the latter setting. Thus, the appointment of (or at least, concurrence in the appointment of) DMC members by the trial sponsor is both natural and appropriate.

Industry and government sponsors have traditionally played different roles in DMC activities. Since there has been some unresolved controversy about whether these differences are appropriate, it is worth describing interactions separately for these two types of sponsor.

7.2.1 Industry sponsors

There are generally continued interactions between the DMC and the sponsor as the study progresses. As described in the previous chapter, the DMC may meet with sponsor representatives during open sessions of DMC meetings to discuss aspects of the trial not requiring the revelation of treatment arm comparisons. Minutes of meetings are generally provided to the sponsor and other components, as described in the previous chapter, although these minutes usually do not discuss treatment arm comparisons. When the DMC recommends a major change in the protocol (early termination of one or more treatment arms, adding new exclusion criteria, modifying the treatment regimen), the DMC may meet with the sponsor to fully review the considerations leading to the recommendation. In cases when the recommendation is to continue the trial but to modify it in some major way, judgment will have to be exercised regarding the extent of information provided to the sponsor. The information must be sufficient to allow the sponsor to decide whether or not to accept the recommendation, but limited enough to permit the comparative results to remain blinded.

We are aware of DMC models that have differed from the above. In some trials, a sponsor representative attends all sessions of the DMC meeting. In other trials, sponsors receive the full DMC reports, including the unblinded comparative data, but do not attend DMC meetings or have any direct interaction with DMC members until the trial is completed (if even then). In such cases, access to the reports may be limited to a few individuals within the sponsoring organization. It is our perception that DMC models in which sponsor representatives have access to interim data, whether by attending the meeting or receiving the report, are being used less and less frequently. The model summarized above and described more fully in Chapter 6, in which the industry sponsor remains blinded to interim comparisons but has the opportunity to direct questions to the DMC and be informed of general trial issues, has an important advantage. It permits the sponsor to maintain involvement while limiting any opportunity for decision-making based on other than scientific grounds (or the perception of such).

7.2.2 Government sponsors

In trials sponsored by federal agencies such as the National Institutes of Health and the Division of Veterans Affairs, agency representatives have typically had regular involvement with confidential DMC operations, in contrast to industry sponsors. They frequently serve as *ex officio* or even full members of DMCs, or as executive secretaries responsible for meeting coordination, taking and/or circulation of minutes, etc. The NIH Policy for Data and Safety Monitoring, issued in June 1998, makes it clear that the sponsoring institute may choose to delegate the data monitoring activities entirely to an outside group but must ensure that an appropriate monitoring system is in place (National Institutes of Health, 1998). The NHLBI policy for data and safety monitoring boards specifies that an institute representative will serve as executive secretary of all DMCs appointed by the NHLBI and in that capacity will be an *ex officio* member (National Heart, Lung and Blood Institute, 2000). National Cancer Institute cooperative oncology group DMC policy calls for an NCI physician and statistician to serve as *ex officio* members of all cooperative group DMCs (National Cancer Institute, 1999).

Despite these guidelines and the history of government sponsor participation in DMCs that preceded the guidelines, some have begun to question these practices (Packer *et al.*, 2001). Representatives of government funding agencies will have to approve decisions about interim changes in the study protocol, judgments about which cannot help but be affected by knowledge of interim data. In addition, concerns have been raised that government sponsors face some of the same conflict-of-interest issues that arise for industry sponsors. Government agencies may perceive that future funding may depend on the success of their clinical trials programs. Staff who have developed the idea for a trial, obtained funding to carry it out, and have been intimately involved in its planning and conduct, may have a great deal of professional investment in the trial and may therefore be overly enthusiastic about early positive results or overly reluctant to act on safety or futility concerns. In one case, cited earlier in Chapter 3, some DMC members publicly protested what they perceived as an overly directive role taken by staff of the sponsoring institution (Strandness, 1995; Imparato, 1996). Although these types of conflict are more subtle, and perhaps of lesser concern than the more direct and personal financial conflicts faced by industry sponsor staff, they may need to be given greater consideration than they have received thus far.

7.3 STUDY STEERING COMMITTEE/PRINCIPAL INVESTIGATOR

Not all clinical trials have steering committees; in those that do, the steering committee (SC) represents the scientific leadership of the trial and, as such, shares some responsibilities with the DMC. (In trials without a formal SC, the principal investigator (PI) may take on the scientific leadership role; in some trials that do have an SC, the PI may represent the SC in interactions with the DMC.) The SC/PI will generally have had the major role in developing the study protocol. A DMC, whether officially required to approve the protocol or not, must at least implicitly accept the study design as appropriate and valid in order to monitor the study according to the plan that has been laid out by the SC/PI. Any concerns raised by the DMC regarding the study design and planned procedures will require interaction with the SC/PI to discuss any changes the DMC views as desirable. Recommended changes in the study protocol are perhaps most likely to occur at the beginning of the trial, but issues may arise during the trial that suggest the need for changes as well. Minutes of DMC meetings, absent any treatment arm comparisons, might be shared with the SC/PI as they are with the sponsor.

Because of the shared responsibilities for trial conduct and oversight of trial quality, it is important for the SC/PI to maintain regular interaction with the DMC. This might take place most naturally at open sessions of DMC meetings (see Chapter 6). One or more SC representatives should generally attend DMC meetings to participate in open sessions and to be available to clarify issues, if necessary, during the DMC closed session deliberations.

At any time during the trial that the DMC recommends major changes, these recommendations are generally presented to the SC/PI as well as the sponsor; while the ultimate decision will in most cases be the sponsor's, the sponsor will want the trial's scientific leadership to participate in the decision-making process. Discussions of such recommendations will be facilitated if SC/PI and sponsor representatives attend the open sessions of DMC meetings and are available to discuss any recommendations directly with the DMC following the closed session.

7.4 STUDY INVESTIGATORS

The DMC usually does not interact directly with study investigators (other than a PI, as discussed in the previous section). Study investigators may receive copies of minutes of DMC meetings if the SC/PI or sponsor chooses to circulate them; more commonly, they may see summaries of any DMC recommendations in communications from the trial sponsor. Any changes in study procedures recommended by the DMC and implemented by the sponsor will of course need to be explained to the study investigators, but this is generally the responsibility of the sponsor and/or the SC/PI.

7.5 TRIAL STATISTICIANS AND STATISTICAL CENTERS

Traditionally, DMCs have had close ties to the statistician who was involved in the design of the trial and, typically, served as a member of the trial SC (if there is one). Often, this trial statistician has prepared the interim analyses reviewed by the DMC at its meetings and, in turn, has attended the DMC meetings to present the interim analyses and participate in the discussion. In many cases, the statistical coordinating center for a trial has operated under the direction of the trial sponsor or the PI.

There has been increasing recognition that the trial statistician as just described faces inherent conflicts. As the individual preparing unblinded interim reports and discussing these reports with the DMC, the statistician clearly must be aware of the unblinded interim results. This knowledge may place the statistician in the difficult position of working with a blinded SC/PI and/or sponsor to make design changes during the trial in response to new information external to the trial, while knowing the potential impact of such changes on the study results.

For example, suppose variable X is the primary endpoint of a study, variable Y is a major secondary endpoint, and both X and Y are important clinical outcomes. Suppose that part-way through the trial, data emerge from a related trial that suggest that the treatment being investigated may have minimal effect on variable X but a strong positive effect on variable Y. The trial leadership may at that point, when they are still blinded to interim results, wish to modify the protocol of the ongoing trial and designate Y as the primary endpoint. The protocol change would be needed to reduce concerns about multiplicity at the end of the trial, and thus protect the validity of inferences drawn from the trial results. If this change is made without knowledge of the interim data, concerns about the interpretability of trial results would be minimized.

If one member of the SC, however, does have access to the unblinded interim data, the picture is changed. It will be difficult for the trial statistician to participate neutrally in a discussion of whether to change the primary endpoint if the statistician knows that the decision will markedly change the likelihood of the trial's ultimately having a positive result. If the decision does lead to a more favorable trial outcome, it will be even more difficult to persuade others that the information held by the statistician played no part in the decision to change the protocol.

7.5.1 The independent statistical center

Such considerations have led to the growing use of an 'independent' statistical center for clinical trials monitored by a DMC (Fisher *et al.*, 2001). All unblinded interim analyses would be prepared at this center, and a statistician from the center would be the one to present the interim analysis to the DMC. This statistician would have had no major role in the design of the study, and would have no routine ongoing interaction with the trial leadership. Thus, the trial structure would include three key statistical components: the primary trial statistician, who is responsible for the statistical aspects of the study design, monitoring the conduct of the study in conjunction with other members of the SC, planning the interim and final analyses, and conducting the final analyses; the independent statistician, who is responsible for carrying out the interim analyses designated by the primary trial statistician, and presenting these analyses to the DMC at regular intervals; and the statistician member(s) of the DMC. Table 7.1 summarizes the potential responsibilities of each of these three roles. It is important to note that this structure provides substantial flexibility. For example, data management and quality control functions could be handled by the primary trial statistician as long as treatment codes were excluded from the data files; or these functions could be handled by the independent statistician.

In trials with an independent statistical center that is separate from the data management center, as noted above, it can be useful for study investigators to send reports of primary and secondary events and/or serious adverse events separately to the independent statistician as well as to the data management center. This process achieves two important goals. First, it establishes an independent channel to verify data, which will later come through the normal data management process. Second, it provides the DMC with an 'up-to-date' accounting of major events at their scheduled meetings where data reports are typically based on a data file created 1–2 months earlier. These independent and up-to-date data can be very helpful to a DMC considering a protocol modification or early termination due to benefit, harm or futility. DMCs are usually uneasy about making recommendations based on data that are not reasonably up-to-date so that data 'in the pipeline' could weaken or even overturn their recommendation. This approach has been used to advantage in several large cardiovascular trials; the cost of such an independent endpoint channel is not great and the benefits can be substantial.

While the concept of the independent statistical center is relatively new, it has the strong advantage of providing flexibility for making needed changes in the study while minimizing any sharing of information that could potentially bias

Table 7.1 Potential roles of statistical components of clinical trials

Function	Primary trial statistician	Independent statistician	DMC statistician
Serves on SC	×		
Involved in design of study	×		
Plans statistical analysis	×		
Helps plan and oversee data management	×	×	
Helps plan and oversee quality control efforts	×	×	
Prepares template for interim reports to DMC	×	×	×
Performs interim analyses and presents to DMC		×	
Has access to data unblinded by treatment code		×	×
Evaluates and interprets interim analyses		×	×
Involved in making changes to protocol before and during the study	×		×
Statistical author on study manuscript(s)	×	×	

the outcome. The disadvantage, of course, is that it adds an extra complexity to the trial administration, and almost surely extra expense. Nevertheless, in our view the separation of the interim analysis function from the study design and leadership function provides important protections of trial integrity and should be considered whenever possible.

It should be noted once again that eliminating all potential conflicts of interest is never possible. Whatever the arrangement for the statistical center, whether it is independent of the trial leadership and sponsor or not, one cannot avoid the problem that a study that stops early may produce less revenue, or proportionally greater revenue, for those who are managing the study data, depending on what sort of funding arrangement was made initially. It is incumbent on all those involved in organizing, managing and conducting clinical trials to consider the potential conflicts and develop strategies to minimize their impact on the study results and interpretation.

7.5.2 Ensuring optimal data presentations

At the beginning of the trial, before data are available for analysis, it is good practice to prepare a plan for interim analyses, including table mock-ups, and present these to the DMC for their consideration. This may be a joint effort of the primary trial statistician, who is responsible for the study's analytical plan, and the independent statistician, who will be preparing and presenting the analyses. It is important for the DMC to have input into the data presentations it is asked to review; if the DMC has preferences regarding data formats, specific tabulations and analyses to be included, graphical approaches, etc., these should be accounted for by the statistical center if at all feasible. As the trial progresses, the DMC may request that supplemental analyses be performed by the independent statistician. On occasion, DMC members may note inconsistencies or other problems in the reports (for example, excessive missing data, imbalances in important prognostic factors, delays in reporting) that they ask the statistical center to address. The DMC and the statistical center share responsibility, to some degree, for the validity and accuracy of the data analyses on which trial decisions are based.

7.6 INSTITUTIONAL REVIEW BOARDS

Institutional review boards (IRBs) review protocols for all studies carried out at that institution, and determine whether they are appropriate, both ethically and scientifically, for the institution's population of potential study participants. Any study being monitored by a DMC will have been reviewed by at least one and usually multiple IRBs. Investigators in each institution are required to report aspects of study progress to their IRBs at regular intervals, or in real time should an event occur that might impact on the IRBs' belief that the study is an appropriate one to carry out. For example, serious adverse events are routinely reported to IRBs.

It is becoming widely recognized that IRBs are limited in their ability to perform meaningful interim reviews of ongoing studies (Department of Health and Human Services 1998; Burman *et al.*, 2001; Morse *et al.*, 2001). Some IRB members have expressed frustration about their responsibility to review what may be a substantial number of adverse events, many occurring in multicenter trials where a single IRB has no information about occurrences in other institutions. With a large number of trials to review in many institutions, and information coming in on all of them (mostly without information about treatment assignment), it is becoming clear that the responsibility for ensuring the continued appropriateness and safety of an ongoing trial is more sensibly that of the DMC (when there is one) than the IRB, particularly for multicenter trials. It has been suggested that a useful approach might be to have the study sponsor or SC/PI circulate the key recommendations of DMC meetings (without providing unblinded and/or comparative data) to all IRBs for which the institution is participating in the trial. This document would provide to the IRB the assurance that a knowledgeable group had performed a thorough review of the interim data and determined that the trial should continue as planned, or made recommendations for changes. The IRBs' review of this report, which would not in most cases reveal treatment group comparisons, could be construed as satisfying the requirement that IRBs monitor the progress of the trial. The NIH has taken this approach for trials under its sponsorship (National Institutes of Health, 1999).

It may be worth noting that since not all trials have DMCs (nor should they), the circulation of DMC minutes to IRBs during trials for which a DMC has been established will not be a complete solution to the difficulties being faced by IRBs.

7.7 REGULATORY AGENCIES

In most cases there is no need for a DMC to interact with regulatory agencies during the course of a trial. In some particularly high-profile trials, regulatory personnel may be invited to and choose to attend open sessions for the DMC meetings. Such interactions may be valuable when it is known that rapid action by the regulatory agency will be expected should the trial be positive, as has been the case for some trials for treatments of HIV infection. As noted in Chapter 4, the regulatory review of data from a trial is most demonstrably objective when the regulatory scientists charged with the review have not been part of the unblinded monitoring process.

Nevertheless, as with so many other 'general principles' discussed in this book, there may be exceptions. There are circumstances in which a DMC may feel the need to contact or consult with the regulatory agency regarding certain findings, and there are circumstances in which the agency may feel the need to approach a DMC. For example, a DMC might notice some unusual toxicities that appear related to a concomitant therapy used by many trial participants and wish to consult with regulatory agency staff before making any recommendation regarding use of this therapy in the trial.

A case in which the FDA approached a DMC, more fully discussed in Chapter 10, came about when the FDA was considering rapid approval of a new agent to treat HIV infection on the basis of results from uncontrolled studies on markers of immunological function. At this time, a DMC overseeing HIV trials for the National Institutes of Health was monitoring several major randomized trials of the same agent. These trials were primarily intended to evaluate clinical effects but also collected the marker data as important secondary information. The FDA believed that rapid approval of new agents was extremely important, since available treatments for the disease were then quite limited. The Agency was concerned, however, about proceeding on the basis of a relatively small data set, when substantial additional data were potentially available in the ongoing trials. Because the trials were near completion, the FDA asked the DMC to concur that the release of the marker data to the FDA by the study statistical center would be acceptable and would not threaten the ability of the trials to reach valid conclusions. Examples of other circumstances in which the FDA has interacted with a DMC are provided in Chapter 10.

Proposals for direct interaction between a DMC and regulatory agency personnel should be made through the study sponsor. The sponsor (and the SC, if there is one) should be aware of, and agree to, any such interaction, although in most cases the sponsor itself would not be involved in reviewing the data of interest. Should this interaction result in any major recommendation for changes in the study, the sponsor and SC would then of course be provided access to the information leading to that recommendation for their concurrence, where this access will be sufficiently limited to permit the comparative results to remain blinded.

7.8 STUDY PARTICIPANTS AND/OR ADVOCACY GROUPS

Study participants clearly are a critical component of any clinical trial, but direct interaction between a DMC and study participants would be rare. One possible circumstance in which such interaction might be beneficial would be a high-profile trial in which participants in the study and/or those considering enrolling in the study wanted to hear directly from the DMC about the monitoring process. In such a case the sponsor might be able to arrange for discussion between one or more DMC members and participant representatives; or a DMC member might make a presentation at a meeting of trial participants and/or potential participants. In general, however, interactions with participants and advocacy groups would be in the purview of the study investigators, SC/PI and sponsor, and not with the DMC.

It is important, however, that potential study participants understand that an expert committee will regularly monitor the trial and that changes in the trial may be implemented based on this review. This information should be part of the informed consent process. In some cases, the DMC may identify unanticipated concerns that are considered insufficient to terminate the trial but important enough to warrant bringing the new information to the attention of trial participants. The DMC may recommend that the sponsor send letters to all study participants, or that participants be brought in for discussion of the new findings, as was done in the HERS trial (Hulley *et al.*, 1998). It may even recommend that study participants undergo a renewed informed consent process to ensure that they remain willing to continue in the study.

7.9 OTHER DATA MONITORING COMMITTEES

Sometimes multiple studies of an investigational agent are ongoing simultaneously. When such studies are monitored by separate DMCs a natural question arises: should the DMCs share accumulating information? A clear advantage of doing so is that it could permit more rapid identification of emerging safety concerns and/or establishment of efficacy. It has even been proposed that, to ensure the most rapid identification of true treatment effects, trial DMCs should base their interim reviews not just on the data from the trial they are monitoring but on a meta-analysis of all similar trials that had been performed, incorporating interim data from the current trial or any other ongoing trial (Chalmers and Lau, 1996).

The disadvantage of sharing interim data among DMCs monitoring related trials is that the trials can no longer be considered entirely independent experiments (Dixon and Lagakos, 2000). In addition, sharing data imposes an additional logistical burden on the trial organizers and an additional review burden on each DMC. In most trial circumstances, when the trial proceeds as planned and no unanticipated concerns rise to the level of requiring some action, these additional burdens are unnecessary.

We believe, however, that occasional and judicious sharing of data among DMCs can be quite valuable. When a DMC finds itself with a very difficult decision to make – for example, possibly recommending that a trial be stopped because of emerging safety concerns even while the interim efficacy data appear quite promising – and interim data from another ongoing trial might substantially reduce the risk of either needlessly stopping the trial or continuing to subject participants to undue risk, it is appropriate for the DMC to seek access to such data (DeMets, 2000).

A detailed example of such a situation is given in Chapter 5. In that case, a DMC observing a difficult-to-explain increased risk on one of two different placebo arms in a trial was provided reassuring data from a similar trial and was therefore comfortable in allowing the trial to proceed. Had the other data not been available, the DMC might have felt it necessary to recommend suspension or closure of the trial it was monitoring. That action would have prevented the trial from answering the questions it had been designed to address, with major adverse consequences: financial consequences both to the pharmaceutical manufacturers whose products were being evaluated and to the government sponsor who would have had to design and implement a new trial to study the issues of interest; but, more importantly, public health consequences to the affected community whose chances of survival could have been enhanced by the information to be gained from the trial. Additionally, since other related trials also were ongoing, those trials could also have been adversely impacted by the early closure of this trial.

Another case of DMC data sharing was initiated by the FDA. A sponsor notified the FDA that the director of a coordinating center for an ongoing trial had provided him with unblinded interim data despite a prior agreement that only the independent DMC would see such data. The coordinating center director did so despite this agreement because of his concerns that an important safety issue was emerging of which the sponsor needed to be aware. The sponsor called the FDA, not being sure what to do about this unsolicited information. FDA staff, aware that a similar trial of the sponsor's product was under way in Europe, recommended that the sponsor ask the DMCs for the two trials to discuss between themselves the data in both trials bearing on the safety issues noted in the first trial. If the concerns were evident only in the first trial, this might provide some reassurance and allow both trials to continue; but if similar concerns were developing in the second trial as well, it might be necessary to terminate both trials. This approach precluded any further unblinding of data to the sponsor or to the FDA. This example illustrates the importance, from the beginning, of having all parties buy into the procedures to be followed. Although there had been an agreement that the sponsor would remain blinded to interim data, the director of the coordinating center became uncomfortable with this agreement. This experience placed the sponsor in the problematic position of having knowledge of interim data at a time when he might have had to make decisions about the trial that could no longer have been made without being influenced by the existing data.

It must be remembered that similar trials may still be different in potentially important ways, so that there will always be some uncertainty about the relevance of the data from one trial population to another. Safety issues arising in one trial of a new agent but not a similar trial might, for example, be attributable to (possibly unrecognized) differences in management practices between investigators in the two trials rather than to the treatment under study. Thus, sharing of interim data is not a 'foolproof' approach to increasing the chances of taking the optimal action. In general, if a DMC believes that it can confidently recommend, on the basis of the data at hand, whether to continue or to make a major change, it should proceed. Seeking data from another ongoing trial is advocated only in cases where the DMC is unsure about whether to make a major recommendation, such as terminating the trial for safety reasons.

REFERENCES

Burman WJ, Reves RR, Cohn DL, Schooley RT (2001) Breaking the camel's back: multicenter clinical trials and local institutional review boards. *Annals of Internal Medicine* **134**: 152–157.

Chalmers TC, Lau J (1996) Changes in clinical trials mandated by the advent of meta-analysis. *Statistics in Medicine* **15**: 1263–1268.

DeMets DL (2000) Relationships between data monitoring committees. *Controlled Clinical Trials* **21**: 54–55.

Department of Health and Human Services (1998) *Institutional Review Boards: Their Role in Reviewing Approved Research* (OEI-01-97-00190). Office of Inspector General, DHHS, June.

Dixon DO, Lagakos SW (2000) Should data and safety monitoring boards share confidential interim data? *Controlled Clinical Trials* **21**: 1–6.

Fisher MR, Roecker EB, DeMets DL (2001) The role of an independent statistical analysis center in the industry-modified National Institutes of Health model. *Drug Information Journal* **35**: 115–129.

Hulley S, Grady D, Bush T, Furberg C, Herrington D, Riggs B, Vittinghoff E, for the Heart and Estrogen/Progestin Replacement Study (HERS) Research Group (1998) Randomized trial of estrogen plus progestin for secondary prevention of coronary heart disease in postmenopausal women. *Journal of the American Medical Association* **280**: 605–613.

Imparato AM (1996) Regarding 'What you didn't know about NASCET'. *Journal of Vascular Surgery* **23**: 182–183.

National Cancer Institute (1999) *Policy of the National Cancer Institute for Data and Safety Monitoring of Clinical Trials.* http://deainfo.nci.nih.gov/grantspolicies/datasafety.htm.

National Heart, Lung and Blood Institute (2000) *Responsibilities of Data and Safety Monitoring Boards (DSMBs) Appointed by the NHLBI.* http://www.nhlbi.nih.gov/funding/policies/dsmb˙inst.htm.

Morse MA, Califf RM, Sugarman J (2001) Monitoring and ensuring safety during clinical research. *Journal of the American Medical Association* **285**: 1201–1205.

National Institutes of Health (1998) NIH policy for data and safety monitoring. *NIH Guide*, June 10. http://grants.nih.gov/grants/guide/notice-files/not98-084.html.

National Institutes of Health (1999) Guidance on reporting adverse events to institutional review boards for NIH-supported multicenter clinical trials. *NIH Guide*, June 11. http://grants.nih.gov/grants/guide/notice-files/not99-107.html.

Packer M, Wittes J, Stump D (2001) Terms of reference for data and safety monitoring committees. *American Heart Journal* **141**: 542–547.

Strandness DE (1995) What you did not know about the North American Symptomatic Carotid Endarterectomy Trial. *Journal of Vascular Surgery* **21**: 163–165.

8

Statistical, Philosophical and Ethical Issues in Data Monitoring

Key Points

- Specialized statistical methods are needed for monitoring clinical trials data to differentiate between 'evidence providing reliable conclusions' and 'fluctuations over calendar time that are consistent with random variability'.
- Several statistical approaches have been developed for evaluating and interpreting data at interim time points during a clinical trial.
- Flexibility, in terms of number and timing of interim analyses, can be built into the statistical monitoring plan.
- DMCs must be in agreement with trial sponsors and trial leadership regarding the statistical and other criteria that will guide recommendations for early termination of the trial.

Every data monitoring committee will face a variety of issues in carrying out its responsibilities. These issues can be far-ranging and include interrelated statistical, philosophical and ethical aspects. This chapter will address these aspects individually, although their interrelatedness will be evident, particularly in the examples.

8.1 THE NEED FOR STATISTICAL APPROACHES TO MONITORING ACCUMULATING DATA

The DMC must review accumulating data periodically in order to assess whether an important safety issue has arisen or whether the intervention under study is providing a substantial and convincing beneficial effect earlier than expected.

The required frequency of these reviews depends on the disease and the specific intervention. Most DMCs hold meetings at least annually and many meet two to four times each year. While these interim reviews are necessary, the process of repeatedly evaluating data must be done with caution, especially early in the course of a trial when the number of participants and the numbers of events related to safety and effectiveness are relatively small. A succession or run of a small number of events on one arm of the trial can appear dramatic, but a few events in succession on the other arm would quickly diminish the overall trend.

This ebb and flow of trends for and against the intervention can be demonstrated by one of the treatment arms in the Coronary Drug Project (Coronary Drug Project Research Group, 1975, 1981). Figure 8.1a displays the behavior over time of the estimated treatment effect measured by the hazard ratio – the mortality risk for the clofibrate arm relative to the placebo arm. In this plot, a value less than 1.0 indicates a lower risk of death for clofibrate. As can be seen, the accumulating data fluctuate several times early in the trial, with positive trends emerging and then disappearing, ultimately stabilizing to an indication of little or no treatment effect (see Figure 8.2 for the final mortality curves). This stabilization occurs because the standard error of the estimate of the hazard ratio is inversely proportional to the square root of the number of events accumulated in the trial. As the number of events increases over calendar time, the standard error will decrease, indicating more precision in the estimate of the treatment effect.

In the Coronary Drug Project, suppose one represents the strength of evidence for the treatment effect by a standardized statistic, called the Z-score. Specifically, the Z-score is the statistic measuring the treatment effect divided by its standard error. In a single analysis of data gathered under a null hypothesis of no treatment effect (a hazard ratio of 1.0), the Z-score will be approximately normally distributed with mean 0 and standard deviation 1. Hence, when there is no treatment effect, at any calendar time, we would expect the Z-score to be between −2 and 2 with approximately 95% probability. Figure 8.1b presents the Z-score when calculated continuously over the passage of calendar time in the Coronary Drug Project; a negative Z-score in this example indicates treatment benefit. It can be seen from the figure that the Z-score fluctuated several times early in the trial between trends toward beneficial effects (Z-scores less than −2) and evidence of no effect (Z-scores near zero).

While a Z-score outside the range of −2 to 2 frequently would be interpreted as evidence that the null hypothesis is not true if data are analyzed at only a single calendar time, it is apparent from Figure 8.1b that values outside that range occur more frequently by chance alone when data are analyzed frequently over time. (This will be explored further in section 8.2.1.) The DMC for the Coronary Drug Project no doubt paid attention to the fluctuating trends, yet made a recommendation at each interim analysis to continue the trial. An early recommendation by the DMC to terminate based on this beneficial trend would have led to the incorrect conclusion that clofibrate was effective in reducing

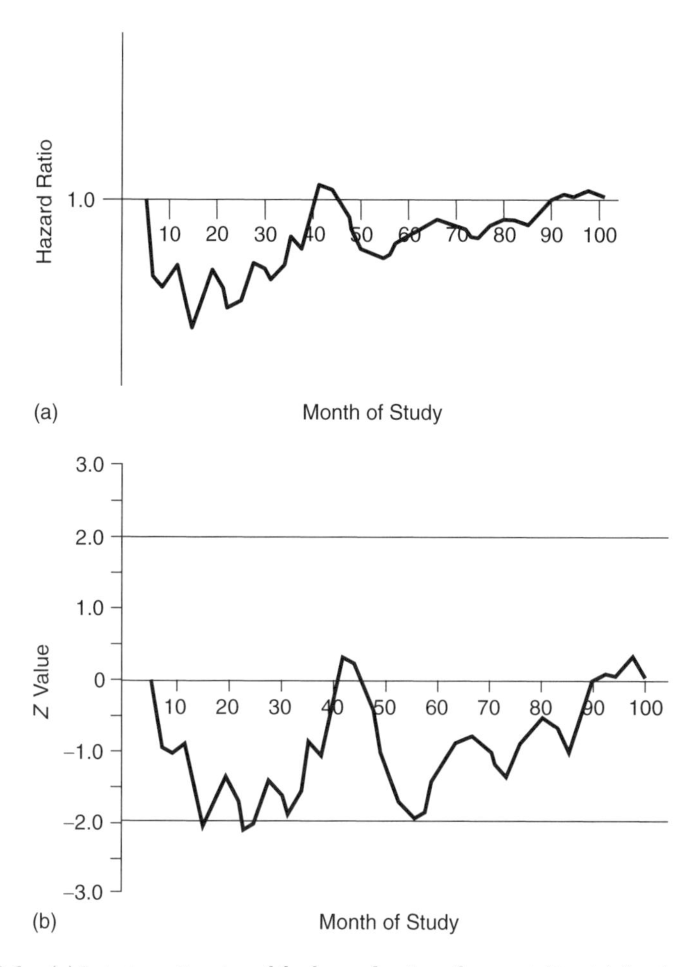

Figure 8.1 (a) Interim estimates of the hazard ratio – the mortality risk for the clofibrate arm relative to the placebo arm – during 100 months of follow-up in the Coronary Drug Project trial. A value less than unity indicates clofibrate superiority, and a value greater than unity indicates placebo superiority. (b) The Z-score for the interim clofibrate–placebo mortality comparison, when calculated continuously during the 100 months of follow-up in the Coronary Drug Project, plotted against the conventional boundaries of ± 1.96. A negative value indicates clofibrate superiority, and a positive value indicates placebo superiority.

mortality in this trial of post-infarction patients. The Coronary Drug Project Research Group (1981) has discussed this particular example further.

While the Coronary Drug Project example demonstrates the need for caution in reviewing early trends, such trends can occasionally be so strong as to provide persuasive evidence of a beneficial or harmful intervention effect. The Cardiac Arrhythmia Suppression Trial (CAST) (CAST II Investigators, 1992; Friedman

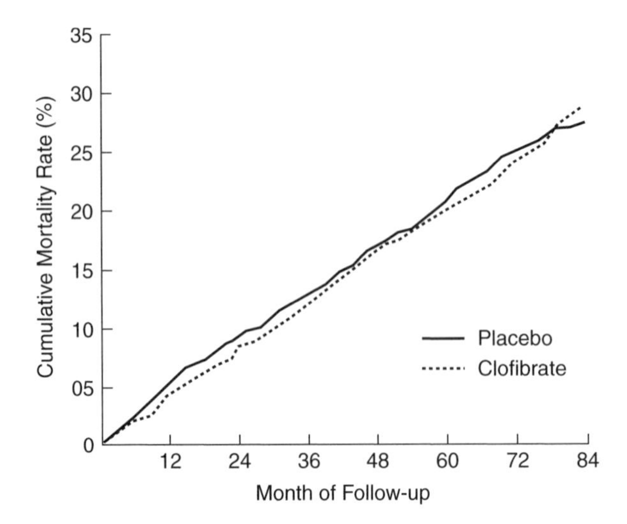

Figure 8.2 Kaplan–Meier mortality curves for the clofibrate and placebo treatment groups in the Coronary Drug Project. From Coronary Drug Project Research Group, 1981, copyright Elsevier Science.

et al., 1993; Task Force of the Working Group on Arrhythmias of the European Society of Cardiology, 1994) was terminated very early due to a harmful effect of drugs that were known to suppress cardiac ventricular rhythm abnormalities. The remarkable feature of the CAST results was that the harmful effect on mortality emerged rapidly and was large in magnitude, despite the fact that the drugs being tested were thought quite likely to be effective in reducing mortality in a population at risk from sudden cardiac death, due to their known effect in suppressing life-threatening premature ventricular beats. However, very early on, when fewer than 15% of the total number of expected deaths had been observed, the Z-score had already reached a magnitude of 3 standard errors, a highly significant and very unexpected result for a therapy expected to be beneficial. The CAST DMC experience has been described in detail (Friedman *et al.*, 1993).

The CDP and CAST experiences illustrate the challenges and dilemmas that a DMC faces in reviewing interim analyses of accumulating results. While statistical methods cannot provide absolute answers to the question of when emerging trends reflect real differences, they can substantially reduce the likelihood of reaching an incorrect decision. Experience has shown the value of statistical monitoring procedures in monitoring outcome data for clinical trials.

8.2 OVERVIEW OF STATISTICAL METHODS

Standard statistical methods include an array of approaches that may be used to assess the evidence provided by interim analyses. While no single approach addresses all of the issues that face a DMC, they do provide useful tools to guide the DMC in its deliberations about emerging trends for safety and benefit. Existing methods can be classified into four categories: group sequential methods for repeated significance testing; the sequential probability ratio test; conditional power or stochastic curtailment; and Bayesian sequential methods.

8.2.1 Group sequential methods

When comparing an experimental intervention against a standard-of-care control in an intrinsically one-sided superiority or non-inferiority setting, using the usual standardized Z-score of 1.96 corresponds to performing a one-sided 0.025 level test and, in turn, to allowing a 2.5% false positive error rate at that analysis. (Note that throughout this chapter we illustrate concepts using one-sided rather than two-sided tests because in our experience most questions studied in clinical trials are fundamentally one-sided. Of course, the standard for strength of evidence corresponding to a 2.5% false positive error rate is achieved whether one is conducting a two-sided 0.05-level test or a one-sided 0.025-level test.) However, as Table 8.1 clearly indicates, when there is no real treatment effect, repeatedly examining accumulating data increases the chances of falsely claiming a beneficial effect if, at each interim analysis, the usual standardized score of 1.96 is used as the criterion for significance. For example, if the 1.96 standard error criterion for treatment effect were used for two interim analyses, the false positive error rate would increase from 0.025 to 0.041, to 0.075 for five interim analyses and to 0.096 for ten interim analyses – see also Armitage *et al.* (1969) or McPherson (1974) for the corresponding impact of interim analyses on the two-sided 0.05-level error rate.

Thus, interim analyses must go beyond just determining whether the standardized difference for any specific analysis is more than 1.96 standard errors or whether the one-sided significance level at a specific point in time is less than 0.025. Group sequential methods have been developed that provide appropriate interpretation of these interim results, accounting for the multiple opportunities to look at the data and draw conclusions. Group sequential procedures use more conservative standardized scores or critical values for the interim analyses than would be used for the final analysis of a study with no interim analyses, in order to achieve the same overall significance level of, say, 0.025 (corresponding to a

Table 8.1 False positive error rate when using a critical value of 1.96 at each interim analysis (i.e., a nominal one-sided $p = 0.025$)

Number*	1	2	3	4	5	10	15	20	25	30
Error Rate**	0.025	0.041	0.053	0.063	0.075	0.096	0.112	0.124	0.133	0.140

*Number of interim analyses.

**Error rate when using the usual standardized Z-score of 1.96 as the criterion for significance and when tests are performed after equal increments of information.

two-sided 0.05 level test) or 0.005 (corresponding to a two-sided 0.01 level test). As will be discussed later in this chapter, in typical one-sided settings one would use these group sequential procedures to formulate an 'upper boundary' to assess strength of evidence for benefit when trends are favorable, and to formulate a 'lower boundary' to assess strength of evidence to rule out benefit or establish harm when trends are unfavorable.

8.2.1.1 *Some group sequential boundaries*

There are many sets of critical values that can achieve the desired control of the overall significance level, each set forming a separate group sequential boundary. Three of these group sequential boundaries are shown in Figure 8.3. These particular boundaries are referred to by the names of the authors who proposed the methods (Haybittle, 1971; Peto *et al.*, 1976; Pocock, 1977; O'Brien and Fleming, 1979). The Pocock approach probably represents the first true group sequential approach, in the sense of being specifically designed for the situation in which interim analyses would be performed at planned regular intervals. (As Pocock himself notes, however, it is primarily of historical interest as more recent approaches incorporate its advantages while eliminating its disadvantages (Ellenberg *et al.*, 1993).)

In Figure 8.3, the standardized statistic or Z-score is plotted on the vertical axis and the fraction of the trial completed on the horizontal axis. The critical values or boundaries for each of the three proposed group sequential methods are plotted over the fraction of the trial information that has been obtained (Lan *et al.*, 1994). (The 'fraction of information' when using a log-rank statistic, for example, would be the proportion of total study events available at the interim analysis.) In this

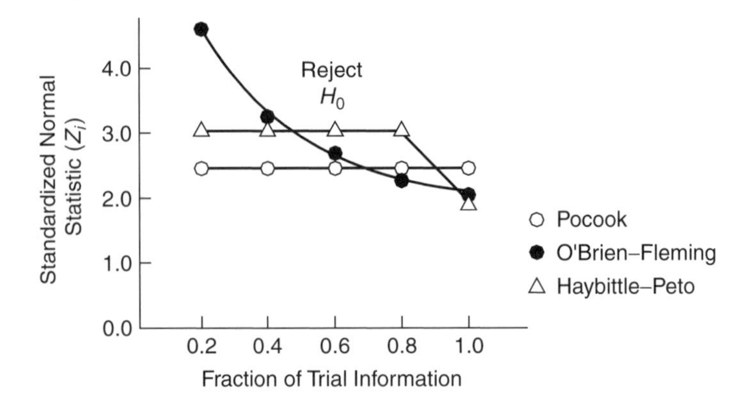

Figure 8.3 Haybittle–Peto, Pocock and O'Brien–Fleming group sequential boundaries for five interim analyses at the one-sided 0.025 significance level. open circle = Pocock; filled circle = O'Brien–Fleming; triangle = Haybittle–Peto, group sequential boundaries.

particular case, the values are given for four interim analyses at 0.2, 0.4, 0.6 and 0.8 of trial information and at the final analysis at the scheduled completion (i.e., at 1.0). For each interim analysis, the summary statistic or Z-score would be plotted and compared to the group sequential boundary that had been selected for a given trial. If the Z-score fell below the boundary, the data would not be considered sufficiently convincing for a finding of benefit and the trial would continue (assuming the decision were to be based only on this particular primary outcome). If the Z-score exceeded the boundary on one of the interim analyses, and all of the other decision factors described later in this chapter were consistent, then the data might well be convincing and the DMC could recommend early termination for benefit. Each of these boundaries controls the overall rate of falsely claiming a beneficial treatment effect at the one-sided 0.025 level (DeMets and Lan, 1994).

If a trial is going to use a group sequential method for guidance in monitoring accumulating data, then one boundary from among the many choices must be selected prior to any review of data. Figure 8.3 shows that the chance of early termination depends on which of the three common boundaries is being used.

The Haybittle boundary (sometimes called the Haybittle–Peto boundary, since these approaches are equivalent) is constant in Figure 8.3 before the time of the final analysis, requiring at least a three standard error treatment difference at any interim analysis before suggesting early termination. Due to the conservativism of this boundary at each interim analysis, the adjustment to the final critical value to obtain an overall one-sided 0.025 significance level is very small. (Fleming *et al.* (1984) explore the adjustment to the conventional 1.96 critical value that is required for statistical significance at the final analysis when using a Haybittle–Peto-type boundary.)

At interim analyses, the Pocock boundary uses a less conservative critical value than the Haybittle–Peto, but requires that the same value be used at all analyses, including the final analysis. A Z-score for a treatment difference that crosses the Haybittle–Peto boundary at an interim analysis would always also cross the Pocock boundary at that analysis (or at one earlier in time). However, in exchange for the ability to use a less conservative critical value at each interim analysis, the Pocock boundary's critical value for the final analysis is much greater than the conventional value of 1.96. Thus, for treatment differences that may emerge later in the trial, the strength of evidence at the final analysis may be sufficient for a conventional critical value but not for the Pocock critical value that is used to control the overall false positive error rate. This could result, for example, in an analysis that achieved a significance level substantially below the nominal level (e.g., 0.01, one-sided, compared with a nominal level of 0.025) not being adequate to reject the null hypothesis. The sample size must also be increased when using Pocock boundaries, in order to achieve the same power as a trial not having interim monitoring. Lan and DeMets (1983) present details of these arguments.

O'Brien and Fleming (1979) proposed what has become one of the most widely used group sequential boundaries. In this case, the boundary values are very extreme early in the trial, when results are still quite unstable. The boundary values become less extreme as the trial progresses, with the critical value at the scheduled final analysis (e.g., 2.04 to maintain a one-sided 0.025 significance level, when conducting analyses at up to five points in time) being close to the conventional critical value (e.g., 1.96). This O'Brien–Fleming boundary has the desirable property of being very conservative early, when one would be skeptical of unstable efficacy and safety results and where data are inadequate to address key model assumptions (such as proportional hazards, or uniformity of effects over important subgroups). The boundary successively relaxes the criteria for significance as information increases and the results become more reliable and less likely to change. This approach is very intuitively appealing and reflects the general philosophy of many who have experience in clinical trial data monitoring. It requires only a negligible increase in sample size since the final critical value is close to the conventional value (Kim and Tsiatis, 1990). If the true treatment effect is of the order of magnitude that was assumed in the protocol's sample size calculations, the trial is not likely to be terminated before 70–80% of the trial information is available on the basis of an O'Brien–Fleming boundary.

8.2.1.2 Group sequential alpha spending functions

The original methodology for group sequential boundaries required that the number and timing of interim analyses be specified in advance. DMCs, however, may require more flexibility as beneficial or harmful trends emerge; waiting 6–12 months for the next look at the data may not be appropriate if there are questions about whether unfavorable safety data may be emerging, for example. Lan and DeMets (1983, 1989a, 1989b) proposed a more flexible implementation of the group sequential boundaries through an 'alpha spending' function. The spending function controls how much of the false positive error (or false negative error when testing to rule out benefit) can be used at each interim analysis as a function of the proportion (t^*) of total information observed. In many applications, t^* may be estimated as the fraction of patients recruited (for dichotomous outcomes) or the fraction of events observed (for time-to-event outcomes) of the total expected. There are alpha spending functions which correspond to or approximate the group sequential boundaries presented in Figure 8.3 as well as many others. For example, an O'Brien–Fleming-type spending function would be

$$\alpha_1(t^*) = 2 - 2\Phi\left[\frac{Z_{1-(\alpha/2)}}{(t^*)^{1/2}}\right],$$

and an approximate Pocock spending function would be

$$\alpha_2(t^*) = \alpha \ln[1 + (\mathrm{e} - 1)t^*].$$

The advantage of the alpha spending function approach is that neither the number nor the exact timing of the interim analyses needs to be specified in advance. Only the particular spending function needs to be specified. The DMC can start out with a particular schedule, but can change the frequency and the timing of the interim analyses as the trends emerge and closer monitoring becomes more critical.

The use of the Lan–DeMets alpha spending approach with an O'Brien–Fleming boundary is illustrated in Table 8.2 and Figure 8.4. Assume four analyses are to be performed, and one wishes to maintain an overall 0.025 false positive error rate. Table 8.2 shows the O'Brien–Fleming guideline when these tests are performed after equal increments of information (i.e., when the proportion of the total information achieved is $t^* = 0.25$, 0.50, 0.75, and 1.0.) The table also indicates the cumulative use of the false positive error, $\alpha(t^*)$, after each proportion, t^*, of the total information is available. Figure 8.4 plots these values, and shows how the Lan–DeMets alpha spending function provides a guideline for the cumulative use of the false positive error, $\alpha(t^*)$, for all values of t^* between 0 and 1.

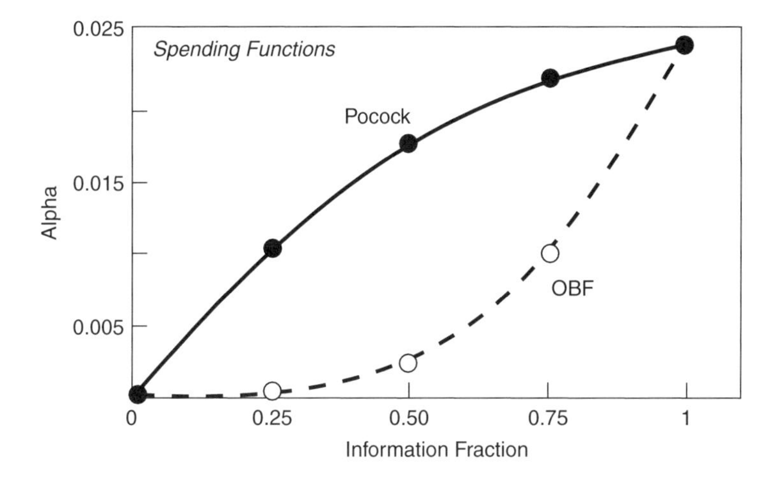

Figure 8.4 Plots of Pocock-type and O'Brien–Fleming-type spending functions for a one-sided 0.025 significance level, for four analyses at 25%, 50%, 75% and 100% of the expected information.

Table 8.2 The O'Brien–Fleming guideline, and cumulative use of the false positive error (maintaining $\alpha = 0.025$, with four analyses after equal increments of information)

Proportion of information, t^*	0.25	0.50	0.75	1.00
Critical value	4.333	2.963	2.359	2.014
Nominal p-value (one-sided)	0.00001	0.0015	0.0091	0.022
Cumulative false positive error, $\alpha(t^*)$	0.00001	0.00153	0.00965	0.025

This alpha spending approach was applied in the Cancer Intergroup 0035 clinical trial (Moertel *et al.*, 1990), introduced in Chapter 5. The study evaluated fluorouracil plus levamisole in the cancer colon adjuvant trial setting, and was designed to follow 973 patients until 500 deaths were observed, with an interim analysis performed after each 125th death. While the first interim analysis was conducted on time (i.e., after 125 deaths had occurred,) the second interim analysis was conducted somewhat later than originally planned, (after 301 deaths had occurred). As can be seen in Figure 8.4, the Lan–DeMets approach indicated that a cumulative (one-sided) $\alpha(0.6) = 0.0038$ could be spent after 301 deaths had occurred. Given that $\alpha(0.25) = 0.00001$ had already been spent at the analysis conducted after the first 125 deaths, straightforward calculations involving properties of the multivariate Gaussian distribution yielded that the proper O'Brien–Fleming monitoring guideline after 301 deaths would be a nominal one-sided 0.0038 level of significance. The fluorouracil plus levamisole regimen had induced an estimated 33% reduction in the death rate, with corresponding one-sided log-rank $p = 0.003$. The regimen had also induced an estimated 40% reduction in the rate of cancer recurrence (one-sided log-rank $p < 0.0001$). Guided by these considerations, the DMC recommended early release of these trial results.

8.2.2 Triangular boundaries

Whitehead (1983, 1994) introduced another method. This method permits unlimited analyses as the trial progresses and as such is called a continuous monitoring procedure. The basis for inference is the test statistic $S_k = Z_k\sqrt{I_k}$ at the kth interim analysis. S_k is referred to as the score statistic; Z_k is known as the standardized statistic and $\sqrt{I_k}$ represents the square root of the accrued information. The sequential boundaries are based on the null hypothesis, $H_0 : \theta = 0$, and the alternative hypothesis $H_A : \theta = \delta$, where θ represents the treatment effect. The boundaries for the test statistic S_k as a function of the information fraction form triangular regions, as shown in Figure 8.5 for a one-sided test. If the test statistic exceeds the upper boundary, the null hypothesis can be rejected. If the test statistic falls below the lower boundary, the trial is terminated and the alternative hypothesis can be rejected. Otherwise, the trial continues until one of the two boundaries is reached. It should be noted that when the total information has been accumulated, whether defined as total sample size or total expected events, the Whitehead boundaries meet and a decision one way or the other must be made.

Using Whitehead's approach, one can calculate the maximum sample size required to achieve a false positive error rate of α and power of $1 - \beta$. The maximum sample size using the triangular boundaries can be shown to be larger than the maximum sample size with group sequential boundaries such as the O'Brien–Fleming. This is to be expected, as the triangular testing approach permits

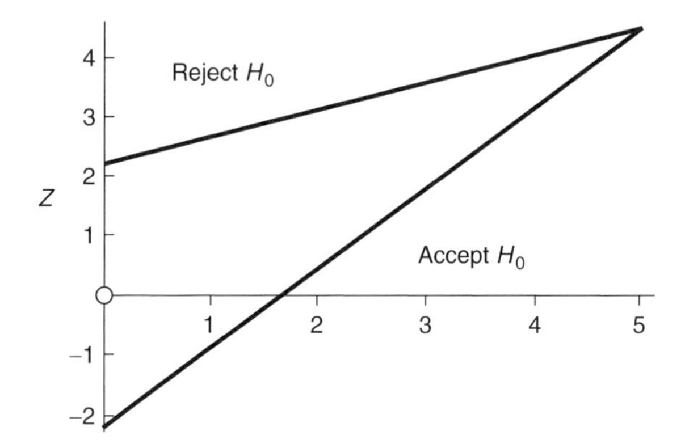

Figure 8.5 A triangular test boundary plotted over information fraction.

interim analysis to be performed far more often than they would be using a group sequential design. A two-sided version of this approach can also be formulated.

8.2.3 Stochastic curtailment

A third data monitoring technique is the method of conditional power or stochastic curtailment (Halperin *et al.*, 1982; Lan *et al.*, 1982; Lan and Wittes, 1988). Most conventional trials are designed to have a high probability of detecting a predefined treatment effect if such an effect truly exists. That probability is called the power of the trial. Typically, the power is set to be from 0.80 to 0.975 for a range of alternatives of interest and the sample size of the study is calculated to achieve that power.

Once a trial is under way and data become available, the probability that a treatment effect will ultimately be detected can be recalculated. An emerging trend in favor of the treatment increases the probability the trial will detect a beneficial effect, while an unfavorable trend decreases the probability of establishing benefit. The term 'conditional power' is often used to describe this evolving probability. The term 'power' is used because it is the probability of claiming a treatment difference at the end of the trial, but it is 'conditional' because it takes into consideration the data already observed that will be a part of the final analysis. This concept was informally used in the Coronary Drug Project (Coronary Drug Project Research Group, 1975), but the statistical methodology has since become more fully developed. Lan and Wittes (1988) provide an especially simple procedure for calculating conditional power for comparing proportions, means and survival curves.

Conditional power can be calculated for an array of scenarios, including a positive beneficial trend, a negative harmful trend or no trend at all. These calculations are most frequently made, however, when interim data are viewed to be unfavorable. In this instance, conditional power represents the probability that the current unfavorable trend would improve sufficiently to yield statistically significant evidence of benefit by the scheduled end of the trial. This probability usually is computed under the assumption that the remainder of the data will be generated from a setting in which the true treatment effect was as large as that originally hypothesized in the study protocol. When an unfavorable trend is observed at the interim analysis, the conditional probability of achieving a statistically significant beneficial effect is much less than the initial power of the trial. If the conditional power is low for a wide range of reasonable assumed treatment effects, including those originally assumed in the study protocol, this might suggest to the DMC that there is little reason to continue the trial since the treatment is unlikely to show benefit. Of course, this conditional power calculation does increase the chance of missing a real benefit (false negative error) since termination eliminates any chance of recovery by the intervention. However, if the conditional power under these scenarios is less than 0.20 relative to the hypothesis for which the trial originally provided 0.85–0.90 power, the increase in the rate of false negative error is negligible. There is no concern with false positive error in this situation since there is no consideration of claiming a positive result.

In the CARS trial (Coumadine Aspirin Reinfarction Study Investigators, 1997), two doses of coumadin were compared to placebo in post heart attack patients, with mortality as the primary outcome. The lower-dose arm was dropped when no emerging trend and a correspondingly low conditional power for treatment benefit was seen. The higher-dose arm continued with a small but positive trend such that the conditional power was somewhat larger than for the low-dose coumadin group. However, this small trend disappeared with further recruitment and follow-up, and the conditional power for this dose became very small as well. At that point, the DMC recommended termination of the higher-dose arm.

In the Beta-Blocker Heart Attack Trial, or BHAT (Beta-Blocker Heart Attack Trial Research Group, 1982; DeMets *et al.*, 1984), the DMC made use of both the O'Brien–Fleming group sequential boundary and stochastic curtailing. With a year of follow-up remaining, the test statistic for the mortality comparison crossed the prespecified O'Brien–Fleming group sequential boundary at the sixth of seven scheduled analyses. The O'Brien–Fleming group sequential boundary and the BHAT results are presented in Figures 8.6 and 8.7. Stochastic curtailment methods were also used to assess the likelihood that the mortality difference would diminish and no longer show statistical significance, should BHAT continue to its scheduled termination. This likelihood was shown to be small, even if no effect was assumed for the remainder of the follow-up. After a thorough discussion of all the relevant factors (see Table 1.1), the DMC recommended that BHAT terminate early. BHAT provides a good example of how group sequential boundaries and stochastic curtailment can be complementary methods for considering early termination recommendations.

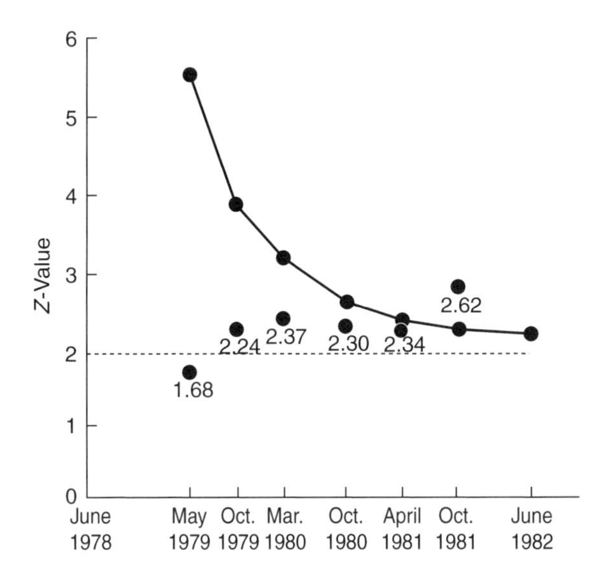

Figure 8.6 Interim mortality results for propranolol vs. placebo comparisons in the Beta-Blocker Heart Attack Trial (BHAT), using a one-sided 0.025 O'Brien–Fleming boundary. From DeMets *et al.*, 1984, copyright Elsevier Science.

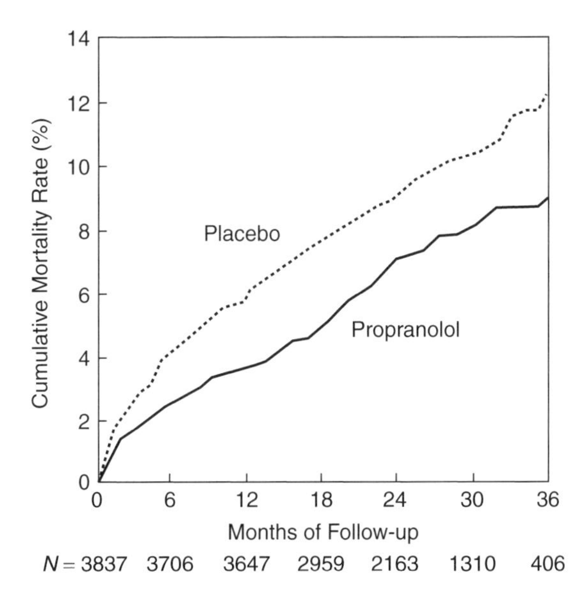

Figure 8.7 Kaplan–Meier mortality curves for the propranolol and placebo arms in the BHAT. From DeMets *et al.*, 1984, copyright Elsevier Science.

8.2.4 Bayesian monitoring

A fourth method for monitoring accumulating data is based on Bayesian methodology (Spiegelhalter *et al.*, 1986; Freedman *et al.*, 1994; Parmar *et al.*, 1994; Fayers *et al.*, 1997). In the Bayesian approach, unknown parameters are considered random and follow probability distributions. The investigators specify a prior distribution or distributions describing the uncertainty in the treatment effect and other relevant parameters. These prior distributions are developed based on previous data and, possibly, beliefs about the treatment effect based on prior data and any other relevant considerations. This specification is quantified through a distribution of possible values and is referred to as the prior distribution. The observed accumulating data are used to modify the prior distribution and produce a posterior distribution, a distribution that reflects the most current information on the treatment effect, taking into account the specified prior (databased or otherwise) as well as the accumulated data. This posterior distribution can then be used to compute a variety of summaries including the predictive probability that the treatment is effective (using the protocol-specified definition of effectiveness) and to compute a credible set of values for the treatment effect θ, with a posterior probability set to some prespecified value (e.g. the posterior probability that the treatment effect is in the region is 0.95 or $1 - 2\varepsilon$ in general). As a specific example, Jennison and Turnbull (1999), take $\varepsilon = 0.025$ so that the updated posterior probability distribution is used to calculate a 'credible set for θ', (θ_L, θ_U), where

$$\Pr(\theta_L < \theta < \theta_U | \text{ the prior distribution and updated data}) = 0.95.$$

This credible set might be used much like a repeated confidence interval (Jennison and Turnbull, 1999) would be used in the group sequential approach. If the credible set excludes θ, one might consider terminating the trial.

One attractive feature of the Bayesian paradigm is that it does not depend on the monitoring schema or monitoring frequency. Still, there is no statistical 'free lunch'. Bayesian analyses of sequentially monitored data are less robust to prior specification than are Bayesian analyses of fixed sample size data sets (Rubin, 1984).

While the Bayesian paradigm has flexibility, it does not necessarily control the false positive error rate. If $\varepsilon = 0.025$, a trial might stop early if

$$\Pr(\theta < 0|\text{data}) < \varepsilon \text{ or } \Pr(\theta > 0|\text{data}) < \varepsilon,$$

which is equivalent to the $1 - 2\varepsilon$ credible set approach (Freedman *et al.*, 1994). This approach to decisions about termination can seriously inflate the false positive rate, much like using a critical value of 1.96 at each interim test. Since the false positive error rate depends on the choice of prior belief for the unknown value of θ, the inflation can be dramatic. Whether or not a Bayesian or a frequentist group sequential paradigm is used in the monitoring process, the control of the rate of reaching false positive conclusions remains important.

The idea of using a Bayesian approach in monitoring clinical trials was introduced by Cornfield (1966). While there is increased interest in Bayesian approaches, and computational tools have made Bayesian analyses more feasible, Bayesian methodology for monitoring trials is not yet widely used. In addition to concerns regarding control of the false positive error rate, specifying a prior distribution for the unknown parameter θ has proven to be challenging. Additional methodological work addresses some of the challenges in using this paradigm (Freedman and Spiegelhalter, 1989; Breslow, 1990; Berry, 1993; Spiegelhalter *et al.*, 1993, 1994; Carlin *et al.*, 1993; Carlin and Louis, 2000; Parmar *et al.*, 2001).

8.2.5 The general approach to sequential stopping boundaries

The above presentation of sequential designs described four approaches in the context of their historical presentation in the statistical literature. Recent work has shown that all of the methods described previously are in fact mere transformations of each other (Emerson 2000; Emerson *et al.*, 2000). Hence, though the Pocock and O'Brien–Fleming boundaries were described above in terms of a standardized Z statistic, they could just as easily be defined on the basis of the partial sum statistic (as was used with the triangular test), the alpha spending function (as described by Lan and DeMets, 1983, and extended by Pampallona *et al.*, 1995), the maximum likelihood estimate of treatment effect (as used by Kittelson and Emerson, 1999, in their description of the unified family of group sequential designs), Bayesian posterior probabilities, or in terms of stochastic curtailment (either conditional power or Bayesian predictive probabilities). The definition of any boundary on one of those scales uniquely determines a boundary on any of the other scales. Families of boundaries can be defined on a variety of scales, and then converted to any other boundary scale of interest.

Because of this equivalence among all of the scales used for the definition of sequential boundaries, which scale is used to derive a particular group sequential design is of less importance than the full operating characteristics desired for the design. Operating characteristics of interest would include the false positive error, the power curve, early conservation when results are unstable, the sample size distribution (which is a function of the true treatment effect), the estimates of treatment effect that would correspond to early stopping, the frequentist inference (p-values and confidence intervals) and Bayesian inference (posterior probabilities of the hypotheses) that would be reported with early stopping, and the stochastic curtailment (conditional power) properties of the design.

8.2.6 Software packages for sequential clinical trial designs

Several statistical software packages are available for formulating statistical monitoring guidelines in clinical trials. Among these are S+SeqTrial (Emerson, 2000; Emerson *et al.*, 2000) from the Insightful Corporation, EaSt 2000 from the Cytel Software Corporation, PEST 4 from the Medical and Pharmaceutical Statistics Research Unit, University of Reading, and LanDeM (Reboussin *et al.*, 2000, from the University of Wisconsin).

8.3 PROTOCOL SPECIFICATION OF THE MONITORING PLAN

The protocol is the blueprint for all critical aspects of the trial, including design, conduct and analysis issues. It should state clearly the objectives of the trial, the primary and secondary outcome measures, and the key safety parameters.

Among the issues that need to be described is the plan for interim monitoring of the accumulating data, including whether or not a DMC is to be established. The most specificity is needed for the description of the statistical methods to be used in monitoring the primary outcome measure, since the study conclusions will rest more heavily on this outcome than any other. In addition, the nature of the statistical monitoring plan has important implications for the design and the sample size, as we have described. Thus, for internal clarity and consistency, some detail is necessary.

However, protocols usually do not describe all operational issues in great detail, in part because experience has shown that each trial is different and flexibility is needed to make adjustments to operational strategies as unanticipated issues emerge. One common approach is to provide the specific detail about the statistical aspects of the interim analysis plan in a separate statistical analysis plan or in the DMC charter (see Chapter 2). While a protocol is the blueprint for the overall trial, the DMC charter often provides the blueprint for the data monitoring procedures (see Appendix A). The statistical analysis plan and the DMC charter can also be appendices to the protocol. This approach keeps the protocol less technical and more readable to the clinical staff who must use this document to guide their implementation of the trial.

8.4 OTHER STATISTICAL CONSIDERATIONS IN MONITORING TRIAL DATA

As discussed in earlier chapters, a variety of factors must be taken into consideration when interpreting interim data. While formal statistical methods do not necessarily exist to account for all of these factors, an understanding and application of basic statistical principles are essential components of the decision-making process when monitoring interim data.

8.4.1 Primary versus secondary endpoints

To avoid multiplicity concerns, most investigators designate a primary endpoint for a clinical trial. Determining what the primary endpoint should be is often not an easy task. The endpoint that is truly most important might occur so infrequently that it could not feasibly serve as the primary basis for inference. There may be multiple important endpoints that the treatment might affect, any one of which could be a basis for encouraging use of the treatment under study. In some cases, a composite endpoint that represents the occurrence of at least one of several different important outcomes is developed and established as the primary endpoint. A DMC should maintain focus on the primary endpoint, particularly for purposes of early termination, because it is often difficult to assess the degree of multiplicity when considering endpoints other than the one designated as primary. Nevertheless, there are cases in which a DMC will put substantial weight on secondary endpoints and may even stop a trial early on the basis of such endpoints.

The Physicians Health Study (PHS) provides a good example of this dilemma. In this study, the primary endpoint had been defined as total cardiovascular mortality, but a number of other endpoints such as fatal and non-fatal myocardial infarction and hemorrhagic stroke were also of major interest and importance. Although total cardiovascular mortality was the defined primary outcome, the observed rate of this outcome at the time the PHS was terminated was only about a tenth of what had been predicted, leading to projections that another ten years of follow-up would be needed to obtain enough fatal events for adequate power. The data on hemorrhagic stroke, another important secondary endpoint, were similarly inadequate. In this case the PHS DMC estimated that too few stroke events would likely be observed, even with a substantial increase in follow-up, to provide information as to whether aspirin would cause hemorrhagic strokes. The combined data on fatal and non-fatal MI, however, showed a very strong benefit from aspirin; the DMC probably could have concluded somewhat earlier in the trial that aspirin reduced the incidence of these events, if that had been the primary endpoint. It is clear that the DMC tried to strike a balance in all these issues (Cairns *et al.*, 1991). Whether the DMC recommended termination too late, too soon, or at the right time will undoubtedly remain a topic for debate. Perspectives depend largely on the importance placed on the secondary outcome of fatal and non-fatal myocardial infarction (MI) as a clinically relevant outcome, weighed against the suggested increase in hemorrhagic strokes even though at a very low rate. What is somewhat instructive is that when the physicians in the trial who were on the placebo arm in the PHS were informed about the aspirin results, including the lack of effect on total cardiovascular mortality, the beneficial effect on fatal and non-fatal MI, as well as the potential bleeding risks, the overwhelming majority (85%) elected to start taking aspirin. Thus, the trial participants, who in this case were especially able to make informed judgments, seem to have agreed with the DMC recommendation and the PHS executive committee's decision. Ethical responsibility to trial participants must remain the top priority for a DMC.

8.4.2 Short-term versus long-term treatment effects

When early data from the clinical trial appear to provide compelling evidence for short-term treatment effects, and yet the duration of patient follow-up is insufficient to assess long-term treatment efficacy and safety, early termination may not be warranted and perhaps could even raise ethical issues. In the Heart and Estrogen/Progestin Replacement Study (HERS) trial (Hulley *et al.*, 1998), early results indicated a statistically non-significant harmful trend for women on hormone replacement therapy (HRT) compared to placebo in the primary outcome of mortality and morbidity. This result appeared to be due mainly to thrombotic events including deep vein thrombosis. As noted in the earlier discussion of this trial (Example 2.12), the DMC recognized the importance of distinguishing between neutral and negative long-term effects on the morbidity/mortality endpoint. Specifically, a neutral result would be consistent with continued widespread use of HRT for several clinical indications including symptom relief. The HERS trial was continued, and achieved average follow-up of 4.1 years at the time of its originally scheduled completion. The final results indicated that the excess of morbidity/mortality events in the first year of treatment was offset by fewer such events in treatment years 4 and 5. Had the trial terminated early, the early unfavorable trends would have led to incorrect conclusions about overall long-term effects.

The relative importance of short-term and long-term results depends on the clinical setting. For example, in studying a treatment that is intended to induce a long-term benefit – anti-inflammatory therapy for rheumatoid arthritis, for example – early short-term benefits (on the primary endpoint) will be viewed with caution, as such short-term effects may readily diminish or even reverse in time. On the other hand, for treatments that are expected to have an acute effect – thrombolytic therapy administered after myocardial infarction, for example – there would be no reason to expect benefits observed early to change with further follow-up. Section 8.5.1 below provides further discussion of issues related to addressing the relative importance of short-term and long-term results in monitoring clinical trials.

8.4.3 Results in subgroups

Whether one is considering interim data or final data, results in subgroups are often difficult to interpret. On the one hand, some variation in estimates of treatment effect among subgroups is expected even when the true treatment effect is uniform across all subgroups of interest. On the other hand, it is usually plausible that a treatment might be more effective (perhaps even much more effective) in certain patient subgroups, or might be helpful to some types of patients and not to others.

When a DMC is reviewing efficacy data, the observation that results are reasonably consistent across subgroups of interest appropriately adds credibility to and confidence in an overall result, while marked inconsistencies in treatment effect across subgroups will usually lead a DMC to be more cautious in interpreting overall results. A DMC should be even more cautious however about drawing conclusions from interim data regarding benefit or harm in particular subgroups.

Subgroup-specific results, known to be unreliable at the time of final analysis, are especially treacherous to interpret at interim analysis.

Two trials in congestive heart failure (CHF) illustrate the dangers of focusing on subgroups. The PRAISE-I (Packer *et al.*, 1996) and PRAISE-II (Packer and the PRAISE-II Study Group, 2000) trials evaluated the effect of the drug amlodipine on mortality and mortality plus CHF hospitalization. In PRAISE-I two subgroups were predefined, according to whether or not the participant's CHF was caused by ischemia. Randomization was stratified by these two subgroups. The predefined hypothesis was that amlodipine might be more effective in ischemia-caused CHF. During the trial the DMC examined results overall, and for each subgroup, at their interim reviews. Overall positive beneficial trends began to emerge for both endpoints, but, contrary to the original hypothesis, nearly all the effect was seen in the non-ischemic subgroup. The DMC did not recommend early termination. In the final analysis, the overall treatment effect for mortality and CHF hospitalization was non-significant ($p = 0.31$) but the treatment by subgroup interaction test for mortality was significant ($p = 0.004$), suggesting the need for separate interpretation of results in each subgroup. The relative risk for mortality was 1.02 for the ischemia subgroup and 0.54 for the non-ischemia subgroup, opposite to the initial expectation. One obvious interpretation of PRAISE-I might have been that amlodipine was effective in reducing mortality in the non-ischemic CHF subgroup of patients. However, because of the inconsistency in the subgroup results and the unexpected nature of this inconsistency, the PRAISE-I investigators chose to be more cautious (the DMC agreed with that interpretation) and suggested that this potentially very important result needed to be confirmed before becoming established as a basis for general treatment recommendations. Thus, they recommended that a second trial be conducted in non-ischemic CHF patients, using a protocol very similar to that used in the first trial. This second trial showed nearly identical event rates in the amlodipine and the placebo control arms (Packer and the PRAISE-II Study Group, 2000), thereby casting doubt on the positive results seen in this subgroup in PRAISE-I.

If a trial is designed to focus on a specific subgroup, and is powered to do so, this risk of false conclusions can be reduced. Such a design was used in the ACTG 019 trial (Volberding *et al.*, 1990) discussed in Chapter 6. This trial examined the effects of zidovidine (AZT) on the progression to AIDS or death in two subgroups defined by whether the person's CD4 count was above or below 500. Those with lower CD4 counts were at a much higher risk of progression to AIDS. Each subgroup was adequately powered to test the effect of AZT on the primary event. During the course of the trial the DMC observed a statistically significant benefit in the higher-risk subgroup, with only a trend to benefit in the other subgroup. Given the substantial side-effects and cost of AZT, it was important to evaluate the risk-to-benefit ratio for the low-risk subgroup as well so that terminating the entire trial was not a desirable option. Thus, the DMC recommended early termination of the higher-risk subgroup and continuation of the trial in the lower-risk subgroup. Ultimately, the trial indicated benefit in both subgroups (Volberding *et al.*, 1994), although later studies suggested that these effects might be limited in duration (Concorde Coordinating Committee, 1994).

On occasion, emerging subgroup results can be spurious due to artifacts in data collection and management. In the NOTT trial (Nocturnal Oxygen Therapy Trial Group, 1980) discussed earlier in Chapter 2, the effect of continuous vs. nocturnal oxygen supplementation on survival and other morbidity measures was tested in patients with advanced chronic obstructive pulmonary disease. This trial was conducted during the late 1970s, prior to development of group sequential designs, so that while the multiplicity problems of repeated testing were recognized, the emergence of nominally significant results at interim analysis was taken as a signal for early termination considerations. During the interim analyses, the DMC examined several subgroups defined by risk factor levels, one being forced expiratory volume in one second (FEV_1). As the trial progressed, the low-FEV_1 subgroup appeared to show a nominally significant effect in favor of continuous oxygen supplementation, without a substantial trend apparent in the higher-FEV_1 subgroup. The DMC requested that the data collection process be examined for completeness. Results from that examination revealed that one or two clinical centers had been tardy in getting all patient data completed and had submitted more mortality data for the nocturnal oxygen treatment group than for the continuous oxygen group. Although the trial was not blinded, there was no evidence that investigator bias caused the reporting delays; the problem seemed simply to be an unintended imbalance in the rate of data collection and reporting. When the data files were updated, the beneficial effects in the low-FEV_1 group were diminished and no longer nominally significant. Thus, the subgroup result at that point in the trial was apparently due to an artifact in data management (DeMets *et al.*, 1982). Had the DMC recommended terminating the trial for the low-FEV_1 subgroup, subsequent data clean-up would have wiped out treatment differences and the opportunity to answer this important question might have been lost. By the time the trial reached its planned completion, a significant treatment benefit favoring continuous oxygen was observed overall, consistent in both high- and low-FEV_1 subgroups (Nocturnal Oxygen Therapy Trial Group, 1980).

8.4.4 Taking external information into account

The DMC must also take into account new information that becomes available as other trials are completed and results presented. When the Norwegian Timolol Trial (Norwegian Multicenter Study Group, 1981) and the Swedish Metoprolol Trials (Hjalmarson *et al.*, 1981) were published, the BHAT trial (Beta-Blocker Heart Attack Trial Research Group, 1982) was about 80% completed in terms of follow-up. All three of these trials were testing beta-blocker drugs in patients who had just survived a heart attack. The general class of beta-blocker drugs was developed to prevent irregular heartbeats and prevent future adverse events such as death or myocardial infarctions. All three trials had mortality as a primary outcome. When published, the Timolol Trial and the Metoprolol Trial showed highly statistically significant benefits for beta blockade. At that time, the BHAT survival comparison had just crossed the O'Brien–Fleming group sequential boundary for benefit as well. The DMC recommendation to terminate the BHAT trial was influenced by the added information that the two other beta-blockers had shown benefit.

Epidemiologic studies have repeatedly shown higher risks of cancer, including lung cancer, in individuals with low serum levels of beta-carotene. Two trials were performed in which the hypothesis was that for individuals at elevated risk of cancer (such as smokers) an increase in levels of serum beta-carotene would lower the incidence of cancer. These trials, the ATBC (Alpha-Tocopherol, Beta-Carotene Cancer Prevention Study Group, 1994) and the Beta-Carotene and Retinol Efficacy Trial (CARET) (Omenn *et al.*, 1996), examined this question in individuals at increased risk of lung cancer due to smoking or asbestos exposure. The DMC for the CARET trial, investigating whether dietary supplementation with beta-carotene could reduce cancer risk, had the results of the ATBC trial available in its deliberations. Contrary to expectation, the ATBC trial had demonstrated a statistically significant increased incidence of lung cancer and lung cancer death in Finnish male smokers treated with high doses of beta-carotene. In the CARET trial, a similar negative trend emerged. The ATBC results, coupled with the strong emerging negative trend, led to the early termination of the CARET trial.

The DMC should not take action on the basis of new results until these have been fully presented, discussed and reviewed by peers as well as by the committee. The new results may not be as fully relevant or as consistent and clear as in the above examples.

8.5 ETHICAL CONSIDERATIONS

The study sponsor and investigators make ethical commitments to the trial's participants that the study will not continue longer than necessary to convincingly establish treatment benefit. Furthermore, a trial with early unfavorable trends will not continue beyond the point where harm can be distinguished from neutrality or, in some cases, will not continue longer than necessary to convincingly rule out treatment benefit. The DMC bears this ethical responsibility on behalf of the investigators.

8.5.1 Early termination philosophies

Before a trial begins, the DMC should have a clear sense of the trial organizers' philosophy regarding early termination in this instance, and the statistical methods providing guidance in the interim evaluation should be consistent with that philosophy. At least three issues should be addressed. First, what magnitude of estimated treatment difference, and over what period of time, would be necessary before a beneficial trend would be sufficiently convincing to warrant early termination? Second, should the same level of evidence be required for a negative trend as for a positive trend before recommending early termination? Third, for a trial with no apparent trend, should the study continue to the scheduled termination?

These considerations are very important because there are major differences in early termination philosophies among clinical trialists. Some investigators take the view that, at least for some trials, the objective should be to produce results that are persuasive enough to effect changes in medical practice (Liberati, 1994). The expectation is that much larger trials and much more precise data are required to meet this goal than the more typical goal of establishing a treatment difference at (two-sided) confidence levels of 0.05 or even 0.01. A DMC monitoring such a trial, in accord with the specified design and objective, would not be considering a recommendation to terminate on the basis of the efficacy outcome unless interim results were substantially more extreme than the already conservative O'Brien–Fleming-type boundaries that are calculated for protection of the usual levels of overall false positive error.

Other investigators are very uncomfortable with the notion that a trial might be continued far longer than might be necessary to persuade most knowledgeable clinical researchers, requiring continued randomization of participants to an inferior treatment regimen for the purpose of persuading practicing clinicians whose openness to changing their practices may be very limited. DMC members must be sure they will be willing to adopt the monitoring philosophy laid out by the trial organizers before agreeing to serve.

8.5.1.1 Responding to early beneficial trends

As recognized in section 8.4.2, determining the optimal length of follow-up can be difficult in a clinical trial having an early beneficial trend. Ideally, evaluating the duration of treatment benefit while continuing to assess possible side-effects or toxicity over a longer period of time would provide the maximum information for clinical use. However, for patients with a life-threatening disease such as heart failure, cancer or advanced HIV/AIDS, strong evidence of substantial short-term therapeutic benefits may be compelling even if it is unknown whether these benefits are sustained over the longer term. Under these circumstances, early termination might be justified to take advantage of this important short-term benefit, with some plan for continued follow-up implemented to identify any serious long-term toxicity. Of course, after the trial has been terminated and patients on the control arm begin to receive the new beneficial treatment, comparisons of the study arms become less meaningful as time progresses. Thus, evaluating long-term side-effects and whether benefit is sustained becomes more difficult.

For patients with a chronic disease such as arthritis, osteoporosis or back pain, the long term effects of the therapy may be of greater importance in evaluating the benefit-to-risk ratio. In this case, a focus on longer-term outcomes may sometimes be justified even in the presence of a strong but short-term beneficial trend. When such diseases are progressive, however, as is the case for arthritis, there will inevitably be a tension between the desire to prevent irreversible disease progression in as many patients as possible, and the desire to understand the long-term effects of the treatment. This tension will have to be resolved on a trial-by-trial basis, and will depend on factors such as the rate of disease progression and the seriousness of its clinical consequences.

When the risk of serious clinical events occurs predominantly in the long term, such as for patients with mild to moderate hypertension, elevated cholesterol levels, retinopathy or early HIV infection, the case for longer-term follow-up can be made even stronger. Here, individuals usually feel healthy and productive. Giving these individuals an intervention to possibly prevent a fatal or irreversible non-fatal event without knowing much about possible longer-term adverse effects and/or having a fuller picture of long-term clinical benefits may not be justified.

The Hypertension Detection and Follow-up Program (HDFP) provides an example (Hypertension Detection and Follow-up Program Cooperative Group, 1979). While cardiologists accepted in the early 1970s that people with severe hypertension should be treated, treatment of mild to moderate hypertension was still uncertain. Such people in general are healthy with normal physical function and thus may not be willing to tolerate side-effects, long-term toxicity or even the inconvenience of taking regular medication that would lessen their quality of life. In the HDFP, patients with mild to moderate hypertension were randomized to agents lowering blood pressure through a combination of available medications or to 'standard care' (less intense intervention) delivered by their private practitioners. Two or three years into the trial, positive mortality trends in favor of the intense strategy began to emerge, yet the trial continued. One reason was that physicians would need to understand the longer-term side-effect profile before being convinced to initiate lifelong therapy for a large patient population. In this case, 'long-term' was five years – the length of follow-up that was established in the original design. After five years, the trial was terminated on schedule and strong mortality benefits were demonstrated with no serious long-term side-effects that would inhibit physicians from treating mild hypertensives. Without the longer-term follow-up, the reluctance to treat may have persisted and hypertensive patients would have continued to be at higher risk than necessary.

The decision to continue the HDFP meant that some patients in the control arm did not get the benefit of the intense strategy until several years after they entered the study. However, had the trial been terminated after only two or three years, those patients' physicians may well have remained reluctant to treat them, not knowing whether there would be negative effects over the longer term. Thus, the HDFP decision may actually have benefited not only the general population of hypertensives but also many of those receiving the control treatment. Similar issues of long-term follow-up have arisen in the use of aspirin in both the primary and secondary prevention of heart attacks (ISIS-2 Collaborative Study Group, 1988; and Steering Committee of the Physicians' Health Study Research Group, 1989). In these cases, differing perspectives remain about whether trial termination occurred too early, too late or at the appropriate time.

In difficult cases such as these, the potential risks and benefits to patients in the trial must be weighed against the potential risks and benefits to the entire population for whom the treatment being assessed would be indicated. These benefits for the broader population might not be accrued if the trial provided insufficiently definitive information such that treating physicians were not persuaded of the benefit of the intervention. Balancing these factors is difficult, and decisions are often controversial; opinions vary on the optimal balance between these needs. It therefore is important for the DMC to establish in early discussions with the trial leadership the desired approach to early stopping considerations. The particular sequential boundaries selected for monitoring should reflect the philosophy of the trial organizers in this regard; DMC members should ensure that they are comfortable with the approach selected.

8.5.1.2 Responding to early unfavorable trends

The issues surrounding early unfavorable trends are even more complicated (DeMets *et al.*, 1999). Such trends can fluctuate as much as positive trends. In the Diabetes Control and Complications Trial (DCCT) (Diabetes Control and Complications Trial Research Group, 1993), an early negative trend reversed itself and a very strong positive result demonstrating the benefit of tight glucose control in diabetic patients was ultimately shown. Had the DMC for the DCCT stopped the trial early because of the early negative trend, a very useful and beneficial treatment strategy for diabetic patients would have been missed, resulting in continued morbidity for these patients. Not all negative trends will reverse themselves, however, and making judgments about the likelihood of a reversal is one of the most difficult tasks a DMC can face.

When an unfavorable trend emerges, three criteria should be considered by a DMC as it wrestles with the question of whether trial modification or termination should be recommended. These criteria can be ordered according to increasing strength of evidence. First, are the trends sufficiently unfavorable that there is very little chance of establishing a significant beneficial effect by the completion of the trial? Second, have the negative trends ruled out the smallest treatment effect of clinical interest? Third, are the negative trends sufficiently strong to conclude a harmful effect? While it is not feasible to plan for every contingency, some prior thought is helpful to a DMC as to which of these three criteria should be given the most attention if early results are unfavorable.

The conditional power method can be used to assess whether an early trend is sufficiently unfavorable that reversal to a significant positive trend is very unlikely or nearly impossible. Under those circumstances, the evidence may be sufficient for the DMC to recommend termination. However, trial organizers must determine whether it is necessary to conclusively establish lack of benefit or even establish harm in order to justify terminating a trial having unfavorable trends (DeMets *et al.*, 1999).

The statistical methods described earlier for obtaining upper boundaries relating to early termination for benefit also provide useful guidelines for distinguishing simple random variation from an effect that either rules out benefit or actually establishes harm. For example, the group sequential methods described can allow for symmetric or asymmetric boundaries. Symmetric boundaries would demand the same level of evidence to terminate early and claim either lack of a beneficial effect or establishment of a harmful effect, as would be required to claim establishment of a beneficial effect. Asymmetric boundaries might allow for less evidence for a negative harmful trend before suggesting early termination. Thus some discussion by the study chair, statisticians and the DMC is necessary to establish the magnitude of negative trend that should be tolerated. For example, an O'Brien–Fleming sequential boundary might be used for monitoring for beneficial effects, while a Pocock-type sequential boundary could provide guidance for safety monitoring.

While the conditional power argument only allows a statement of failure to establish benefit, the symmetric or asymmetric boundary approach allows the researchers to rule out benefit for a treatment or, with more extreme results, to establish harm. Figure 8.8 provides examples of lower boundaries, either for establishing lack of a beneficial effect (denoted by open circles) or for establishing a harmful effect (denoted by filled circles). In Figure 8.8, the primary test statistic, denoted by Z, is plotted on the y-axis. Data from the PROMISE trial, showing an early unfavorable trend, are presented in Figure 8.9 to illustrate the usefulness of these boundaries in guiding judgments regarding the strength of evidence provided by early unfavorable trends.

Assume one plans multiple interim analyses, using symmetric O'Brien–Fleming monitoring guidelines. Suppose the effect of treatment is represented by δ, such as the true difference in success rates, the logarithm of the true odds ratio or, in a time-to-event analysis, the logarithm of the true relative risk. Then, the upper boundary in Figure 8.8 provides the guideline for strength of evidence required to establish benefit (i.e., to rule out that $\delta = 0$, in favor of benefit), when one wishes to maintain a one-sided $\alpha = 0.025$ false positive error rate. Further, suppose δ_0 represents the level of therapeutic benefit that is sufficiently important clinically that it should be detected with 0.975 probability. If a trial has 0.975 power to detect $\delta = \delta_0$, the lower boundary to establish lack of a beneficial effect (i.e., ruling out $\delta = \delta_0$), is denoted by open circles. This lower boundary is symmetric with the upper boundary 'in the sense that designing a test with the null and alternative hypotheses interchanged would result in identical boundaries', as

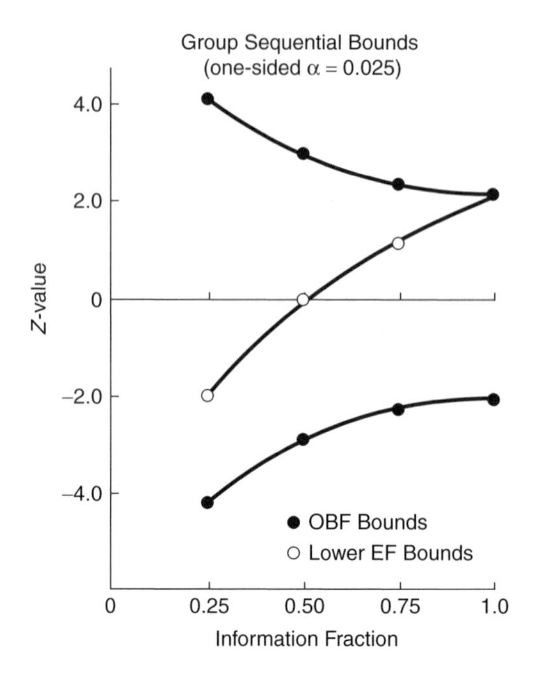

Figure 8.8 A group sequential plan with a 0.025 O'Brien–Fleming-type boundary for benefit (positive), a symmetric O'Brien–Fleming-type boundary for harm (negative) and a symmetric Emerson–Fleming (EF) boundary for lack of benefit.

stated by Emerson and Fleming (1989). In contrast, the lower boundary to establish harm (i.e., ruling out $\delta = 0$, in favor of harm), is denoted by filled circles and is symmetric with the upper boundary through the horizontal line $y = 0$.

Although not illustrated in Figure 8.8, when early trends are unfavorable in a trial that is properly powered, stochastic curtailment criteria would generally yield monitoring criteria for termination that would be similar to the symmetric lower boundary for lack of benefit. However, in underpowered trials having unfavorable trends, the stochastic curtailment criteria for early termination generally would be satisfied earlier than criteria based on the group sequential lower boundary for lack of benefit.

In the CONSENSUS-II trial (Swedberg *et al.*, 1992), the DMC made a recommendation to terminate early with a negative trend that did not statistically significantly establish harm but did rule out a beneficial effect. If the treatment, which was a standard drug delivered in a non-standard bolus, was not going to be substantially better than conventional dosage, it was unlikely to be of much clinical interest. In contrast, the PROMISE (Packer *et al.*, 1991) and VEST (Cohn *et al.*, 1998) trials in congestive heart failure continued with emerging negative trends in mortality in order to distinguish between a small beneficial or neutral effect and a true harmful effect (see Figure 8.9). These trials evaluated a class of

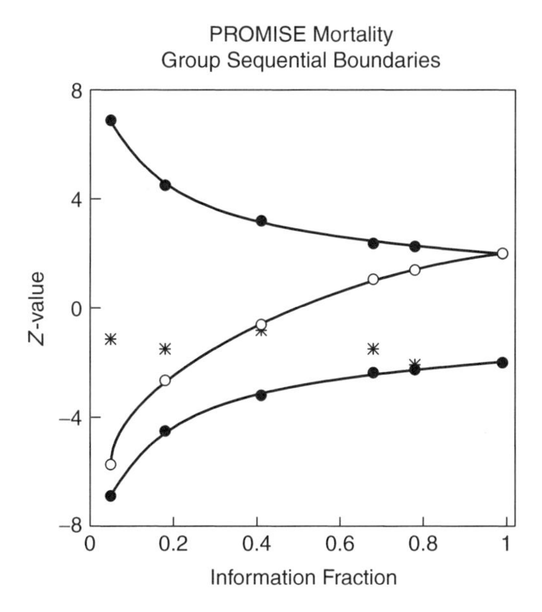

Figure 8.9 A group sequential plan (Figure 8.8) applied to interim mortality comparisons for milrinone vs. placebo arms in the PROMISE trial.

drugs for treatment of heart failure that made the failing heart work harder and allowed patients to exercise longer or feel better. In this situation, it was important to distinguish between a drug that had no beneficial effect on mortality but improved exercise capacity and quality of life, from a drug that had these benefits but increased mortality. Thus, allowing each trial to continue long enough to make this distinction was a difficult, but ethically and scientifically compelling, decision for each DMC. Ultimately in both trials, as illustrated for the PROMISE trial in Figure 8.9, the longer-term mortality data demonstrated a statistically significant harmful effect, despite early quality of life improvements. Not allowing these trials to distinguish between a neutral and a harmful mortality effect would have left physicians and patients in a quandary, resulting in many heart-failure patients, including those in the trials, being subjected to continued exposure to drugs that could decrease their chances of survival. DeMets *et al.* (1999) have discussed this dilemma of the agonizing negative (i.e., harmful) trend, and provide several additional examples of such circumstances.

Most trials comparing a new intervention to a standard of care or a control regimen do not set out to establish that the new intervention is inferior to the control. However, some circumstances may in fact lead to such a consideration. For example, a newly developed drug, biologic, procedure or device may become widely used for an indication before it has been studied in a definitive trial. This was in fact the case with the use of a class of anti-arrhythmic drugs that were known to suppress cardiac ventricular arrhythmias, and were becoming widely used in a broad category of patients having arrhythmias that were not nearly as severe as those in patients who were initially studied. The CAST trial (CAST II Investigators, 1992; Friedman *et al.*, 1993) set out to establish the survival benefit of these drugs in this broader class of patients using a design with a one-sided 0.025 level of significance. The DMC imposed a lower boundary at a 0.025 significance level, not expecting it to be really used. As discussed earlier in this chapter, negative mortality trends began to emerge very early but, because of strong prior belief in the benefits of these drugs, such trends were unlikely to have had an impact on treatment practices. (At the first interim analysis, as noted in Chapter 5, the DMC was actually blinded to direction.) The CAST mortality results rapidly became much more strongly negative, with the Z-score crossing the symmetric lower boundary related to establishing harm. At that point, the DMC was unblinded and recommended that CAST be terminated immediately. The results persuaded the cardiology community that a treatment widely believed to be the 'state of the art' in people at moderate risk of cardiac arrhythmias was, in fact, harmful. Less extreme results might not have been as convincing, and treatment with harmful drugs might have continued.

8.5.1.3 *Responding to unexpected safety concerns*

Statistical methods are least helpful when an unexpected and worrying toxicity profile begins to emerge. In this situation there can be no prespecified statistical plan, since the outcome being assessed was unanticipated. Further, since the number of possible types of adverse events that could be observed and raise concerns is very large, it is also difficult to prespecify any statistical plan for assessing such trends that could control adequately for the multiplicity problems. Nevertheless, monitoring for such problems is a critical responsibility of a DMC. DMC members, in considering such trends, will derive some guidance from a general awareness of the role of coincidence and chance, but will need to rely most heavily on their core knowledge, experience and common sense in considering recommendations based on unexpected safety concerns.

The level of a safety concern that would lead to a DMC recommendation for modification or termination of the study will necessarily vary with the level of benefit being observed. If the treatment appears to be offering a survival benefit, for example, a strong suggestion of a serious and unexpected safety problem might lead to some change in the protocol to reduce the problem (e.g. change in dose, clinical monitoring procedures, use of concomitant medications to prevent or mitigate the problem), while the same magnitude of safety concern might lead to a recommendation to terminate the study if the interim efficacy results were less promising. For this reason, as noted in Chapter 6, when interim monitoring of comparative data is conducted to assess safety issues, efficacy data should also be reviewed in order to enable an informed assessment of the benefit-to-risk profile.

8.5.1.4 Responding when there are no apparent trends

In some trials, no apparent trends of either beneficial or harmful effects emerge as the trial progresses toward its planned conclusion. In such instances, a decision must be made as to whether the investment in participant, physician and fiscal resources, as well as the burden of the trial to patients, remains viable and compelling. For example, if participants are being exposed to an investigational intervention that has some toxicity or invasiveness of administration, the further information on the intervention or the disease itself that might be gained by continuing the trial may not be justified. For many trials, fiscal considerations for both government and industry sponsors may require serious discussion about whether to terminate the trial and allocate limited patient and fiscal resources to more promising interventions. In the CARS trial, two low fixed doses of coumadin were compared to a placebo in a post-myocardial infarction patient population. It was determined that the low- and high-dose arms failed to provide either coagulation or other apparent treatment effects. Because of the potential risk of coumadin, patients had to undergo intensive blood monitoring. Due to this burden and due to the lack of apparent benefit, the trial was terminated. However, for those interventions that do not induce a sizable or serious patient burden, the potential to learn more about the natural history of the disease, or secondary study outcomes, might justify trial continuation.

The DMC charter should discuss these options in sufficient detail to provide the DMC guidance in dealing with these difficult situations should they arise. Prior thought is extremely helpful, since investigator and sponsor input on these issues is difficult to obtain once the trial is under way if confidentiality is to be preserved.

8.5.2 Other ethical considerations

A variety of ethical considerations are very much a part of the data monitoring process, as has already been indicated often in this book. The implicit contract with any patient or individual participating in a clinical trial is that the trial duration will be no longer than necessary to achieve the objectives as stated in the protocol and explained in the consent process. Pursuing other objectives, when those formally defined in the protocol have been adequately addressed, can be ethically troublesome. Patients should not be at risk of exposure to inferior or harmful treatments longer than necessary to establish the benefit-to-risk profile of those treatments. For example, treatments that have been shown to be life-saving or to prevent serious irreversible events should be made available to those patients in the trial as well as to future patients as soon as possible.

In addition to benefit and risk assessments, the DMC has an ethical responsibility to monitor the quality and the viability of the trial. If the DMC recognizes that the design assumptions are no longer plausible, and the trial has no chance to evaluate rigorously the benefit-to-risk ratio, then the trial may serve no useful purpose. In some cases, recruitment may be so seriously behind schedule that the trial cannot be completed in a reasonable time frame or within the period of allocated funding. Data collection may be inadequate in terms of quality or timeliness. Adherence to the study protocol may be very poor and/or the rates of dropout and loss to follow-up may be very high. In these instances, the goals of the trial may not be achievable. Continuation of the trial contradicts the commitment or contract that the investigators made with the patients: that their participation would allow the objectives of the trial to be met. Patients likely would neither initially participate nor wish to continue in a trial that had little chance of adequately evaluating the intervention.

REFERENCES

Alpha-Tocopherol, Beta-Carotene Cancer Prevention Study Group (1994) The effect of vitamin E and beta carotene on the incidence of lung cancer and other cancers in male smokers. *New England Journal of Medicine* **330**: 1029–1035.

Armitage P, McPherson CK, Rowe BC (1969) Repeated significance tests on accumulating data. *Journal of the Royal Statistical Society, Series A* **132**: 235–244.

Berry DA (1993) A case for Bayesianism in clinical trials (with discussion). *Statistics in Medicine* **12**: 1377–1404.

Beta-Blocker Heart Attack Trial Research Group (1982) A randomized trial of propranolol in patients with acute myocardial infarction. I. Mortality results. *Journal of the American Medical Association* **247**: 1707–1714.

Breslow NE (1990) Biostatistics and Bayes (with discussion). *Statistical Science* **5**: 269–298.

Cairns J, Cohen L, Colton T, DeMets DL, Deykin D, Friedman L, Greenwald P, Hutchinson GB, Rosner B (1991) Issues in the early termination of the aspirin component of the Physicians' Health Study. Data Monitoring Board of the Physicians' Health Study. *Annals of Epidemiology* **1**(5): 395–405.

Carlin BP and Louis TA (2000) *Bayes and Empirical Bayes Data Analysis*. Boca Raton, FL: Chapman & Hall/CRC Press.

Carlin BP, Chaloner K, Church T, Louis TA, Matts JP (1993) Bayesian approaches for monitoring clinical trials, with an application to toxoplasmic encephalitis prophylaxis. *The Statistician* **42**: 355–367.

Cohn JN, Goldstein SO, Greenberg BH, Lorell BH, Bourge RC, Jaski BE, Gottlieb SO, McGrew F, DeMets DL, White BG, for the Vesnarinone Trial Investigators (1998) A dose-dependent increase in mortality with vesnarinone among patients with severe heart failure. *New England Journal of Medicine* **339**(25): 1810–1816.

Concorde Coordinating Committee (1994) Concorde: MRC/ANRS randomised double-blind controlled trial of immediate and deferred zidovudine in symptom-free HIV infection. *Lancet* **343**: 871–881.

Cornfield J (1966) A Bayesian test of some classical hypotheses – with applications to sequential clinical trials. *Journal of the American Statistical Association* **61**: 577–594.
Coronary Drug Project Research Group (1975) Clofibrate and niacin in coronary heart disease. *Journal of the American Medical Association* **231**: 360–381.
Coronary Drug Project Research Group (1981) Practical aspects of decision making in clinical trials. The Coronary Drug Project as a case study. *Controlled Clinical Trials* **1**: 363–376.
Coumadine Aspirin Reinfarction Study (CARS) Investigators (1997) Randomized double-blind trial of fixed low dose warfarin with aspirin after myocardial infarction. *Lancet* **350**(9075): 389–396.
DeMets DL and Lan KKG (1994) Interim analyses: The alpha spending function approach. *Statistics in Medicine*, **13**(13/14): 1341–1352.
DeMets DL, Williams GW, Brown BW and the NOTT Research Group (1982) A case report of data monitoring experience: The Nocturnal Oxygen Therapy Trial. *Controlled Clinical Trials* **3**: 113–124.
DeMets DL, Hardy R, Friedman LM and Lan KKG (1984) Statistical aspects of early termination in the Beta-Blocker Heart Attack Trial. *Controlled Clinical Trials* **5**: 362–372.
DeMets DL, Pocock S, Julian DG (1999) The agonizing negative trend in monitoring clinical trials. *Lancet* **354**: 1983–1988.
Diabetes Control and Complications Trial Research Group (1993) The effect of intensive treatment of diabetes on the development and progression of long-term complications in insulin-dependent diabetes mellitus. *New England Journal of Medicine* **329**(14): 977–986.
Echt DS, Liebson PR, Mitchell LB, *et al.* (1991) Mortality and morbidity in patients receiving encainide, flecainide, or placebo. The Cardiac Arrhythmia Suppression Trial. *New England Journal of Medicine* **324**: 781–788.
Ellenberg SS, Geller N, Simon R, Yusuf S (1993) (eds.) Proceedings of 'Practical issues in data monitoring of clinical trials', Bethesda, Maryland, USA, 27–28 January 1992. *Statistics in Medicine* **12**: 415–616.
Emerson SS (2000) S+SeqTrial Technical Overview. Technical Report, Insightful Corporation, Seattle.
Emerson SS, Fleming TR (1989) Symmetric group sequential test designs. *Biometrics* **45**: 905–923.
Emerson SS, Bruce AG, Baldwin K (2000) *S+SeqTrial Users Guide*. Seattle: Insightful Corporation.
Fayers PM, Ashby D, Parmar MKB (1997) Tutorial in biostatistics. Bayesian data monitoring in clinical trials. *Statistics in Medicine* **16**: 1413–1430.
Fleming TR, Harrington DP, O'Brien PC (1984) Designs for group sequential tests. *Controlled Clinical Trials* **5**: 348–361.
Friedman LM, Bristow JD, Hallstrom A, Schron E, Proschan M, Verter J, DeMets DL, Fisch C, Nies AS, Ruskin J, Strauss H, Walters L (1993) Data monitoring in the Cardiac Arrhythmia Suppression Trial. *Online Journal of Current Clinical Trials* Doc. 79, July 31.
Freedman LS, Spiegelhalter DJ (1989) Comparison of Bayesian with group sequential methods for monitoring clinical trials. *Controlled Clinical Trials* **10**(4): 357–367.
Freedman LS, Spiegelhalter DJ, Parmar MK (1994) The what, why and how of Bayesian clinical trials monitoring. *Statistics in Medicine* **13**(13–14): 1371–1383.
Halperin M, Lan KKG, Ware J, Johnson NJ, DeMets DL (1982) An aid to data monitoring in long-term clinical trials. *Controlled Clinical Trials* **3**: 311–323.
Haybittle JL (1971) Repeated assessment of results in clinical trials of cancer treatment. *British Journal of Radiology* **44**: 793–797.
Hjalmarson A, Herlitz J, Malek I, *et al.* (1981) Effect on mortality of metoprolol in acute myocardial infarction: a double-blind randomized trial. *Lancet* **2**: 823–827.
Hulley S, Grady D, Bush T, Furberg C, Herrington D, Riggs B, Vittinghoff E, for the Heart and Estrogen/Progestin Replacement Study (HERS) Research Group (1998) Randomized

trial of estrogen plus progestin for secondary prevention of coronary heart disease in postmenopausal women. *Journal of the American Medical Association* **280**: 605–613.

Hypertension Detection and Follow-up Program Cooperative Group (1979) Five-year findings of the hypertension detection and follow-up program. I. Reduction in mortality of persons with high blood pressure, including mild hypertension. *Journal of the American Medical Association* **242**(23): 2562–2571.

ISIS-2 Collaborative Group (1988) Randomised trial of intravenous streptokinase, oral aspirin, both, or neither among 17,187 cases of suspected acute myocardial infarction: ISIS-2. *Lancet* **2**: 349–360.

Jennison C, Turnbull B (1999) *Group Sequential Methods with Applications to Clinical Trials.* Boca Raton, FL: Chapman & Hall/CRC.

Kim K, Tsiatis AA (1990) Study duration for clinical trials with survival response and early stopping rule. *Biometrics* **46**: 81–92.

Kittelson JM, Emerson SS (1999) A unifying family of group sequential test designs. *Biometrics* **55**: 874–882.

Lan KKG, DeMets DL (1983) Discrete sequential boundaries for clinical trials. *Biometrika* **70**(3): 659–663.

Lan KKG, DeMets DL (1989a) Changing frequency of interim analyses in sequential monitoring. *Biometrics* **45**: 1017–1020.

Lan KKG, DeMets DL (1989b) Group sequential procedures: Calendar versus information time. *Statistics in Medicine* **8**: 1191–1198.

Lan KKG and Wittes J (1988) The *B*-value: A tool for monitoring data. *Biometrics* **44**: 579–585.

Lan KKG, Simon R, Halperin M (1982) Stochastically curtailed tests in long-term clinical trials. *Communications in Statistics C, Sequential Analysis* **1**: 207–219.

Lan KG, Reboussin DM, DeMets DL (1994) Information and information fractions for design and sequential monitoring of clinical trials. *Communications in Statistics: Theory and Methods* **23**: 403–420.

Liberati A (1994) Conclusions. 1. The relationship between clinical trials and clinical practice: the risks of underestimating its complexity. *Statistics in Medicine* **13**: 1485–1491.

McPherson K (1974) Statistics: the problem of examining accumulating data more than once. *New England Journal of Medicine* **290**: 501–502.

Moertel CG, Fleming TR, MacDonald JS, Haller DG, Laurie JA, Goodman PJ, Ungerleider JS, Emerson WA, Tormey DC, Glick JH, Veeder MH and Mailliard JA (1990) Levamisole and fluorouracil for adjuvant therapy of resected colon carcinoma. *New England Journal of Medicine* **322**: 352–358.

Norwegian Multicenter Study Group (1981) Timolol-induced reduction in mortality and reinfarction in patients surviving acute myocardial infarction. *New England Journal of Medicine* **304**: 801–807.

Nocturnal Oxygen Therapy Trial Group (1980) Continuous or nocturnal oxygen therapy in hypoxemic chronic obstructive lung disease – A clinical trial. *Annals of Internal Medicine* **93**(3): 391–398.

O'Brien PC, Fleming TR (1979) A multiple testing procedure for clinical trials. *Biometrics* **35**: 549–556.

Omenn GS, Goodman GE, Thornquist MD, Balmes J, Cullen MR, Glass A, Keogh JP, Meyskens FL Jr, Valanis B, Williams JH Jr, Barnhart S, Hammar S (1996) Effects of a combination of beta carotene and vitamin A on lung cancer and cardiovascular disease. *New England Journal of Medicine* **334**: 1150–1155.

Packer M, Carver JR, Rodeheffer RJ, Ivanhoe RJ, DiBianco R, Zeldis SM, Hendrix GH, Bommer WJ, Elkayam U, Kukin ML, Mallis GI, Sollano JA, Shannon J, Tandon PK, DeMets DL for the PROMISE Study Research Group (1991) Effect of oral milrinone on mortality in severe chronic heart failure. *New England Journal of Medicine* **325**(21): 1468–1475.

Packer M, O'Connor CM, Ghali JK, Pressler ML, Carson PE, Belkin RN, Miller AB, Neuberg GW, Frid D, Wertheimer JH, Cropp AB, DeMets DL, for the Prospective Randomized Amlodipine Survival Evaluation Study Group (1996) Effect of amlodipine on morbidity and mortality in severe chronic heart failure. *New England Journal of Medicine* **335**: 1107–1114.

Packer M and the PRAISE II Study Group (2000) Mortality results for the PRAISE II trial. Paper presented to the American Heart Association Annual Meeting, Anaheim, November.

Pampallona SA, Tsiatis AA, Kim KM (1995) Spending functions for the type I and type II error probabilities of group sequential tests. Technical Report, Department of Biostatistics, Harvard School of Public Health.

Parmar MK, Spiegelhalter DJ, Freedman LS (1994) The CHART trials: Bayesian design and monitoring in practice. CHART Steering Committee. *Statistics in Medicine* **13**(13–14): 1297–1312.

Parmar MK, Griffiths GO, Spiegelhalter DJ, Souhami RL, Altman DG, van der Scheuren E, CHART Steering Committee (2001) Monitoring of large randomized clinical trials: a new approach with Bayesian methods. *Lancet* **358**: 375–381.

Peto R, Pike MC, Armitage P *et al.* (1976) Design and analysis of randomized clinical trials requiring prolonged observations of each patient. I. Introduction and design. *British Journal of Cancer* **34**: 585–612.

Pocock SJ (1977) Group sequential methods in the design and analysis of clinical trials. *Biometrika* **64**: 191–199.

Reboussin DM, DeMets DL, Kim K, Lan KKG (2000) Computations for group sequential boundaries using the Lan–DeMets spending function method. *Controlled Clinical Trials* **21**: 190–207.

Rubin DB (1984) Bayesianly justifiable and relevant frequency calculations for the applied statistician. *Annals of Statistics* **12**: 1151–1172.

Spiegelhalter DJ, Freedman LS, Blackburn PR (1986) Monitoring clinical trials: Conditional or predictive power? *Controlled Clinical Trials* **7**: 8–17.

Spiegelhalter DJ, Freedman LS, Parmar MK (1993) Applying Bayesian ideas in drug development and clinical trials. *Statistics in Medicine* **12**(15–16): 1501–1511.

Spiegelhalter DJ, Freedman LS, Parmar MKB (1994) Bayesian approaches to randomised trials. *Journal of the Royal Statistical Society A* **157**: 357–416.

Steering Committee of the Physicians' Health Study Research Group (1989) Final report on the aspirin component of the ongoing Physician's Health Study. *New England Journal of Medicine* **321**: 129–135.

Swedberg K, Held P, Kjekhus J, Rasmussen K, Ryden L, Wedle H (1992) Effects of early administration of enalapril on mortality in patients with acute myocardial infarction – results of the Cooperative New Scandinavian Enalapril Survival Study II (Consensus II). *New England Journal of Medicine* **327**: 678–684.

Task Force of the Working Group on Arrhythmias of the European Society of Cardiology (1994) The early termination of clinical trials: Causes, consequences, and control. With special reference to trials in the field of arrhythmias and sudden death. *Circulation* **89**: 2892–2907.

Volberding PA, Lagakos SW, Koch MA, Pettinelli C, Myers MW, Booth DK, Balfour HH, Reichman RC, Bartlett JA, Hirsch MS, Murphy RL, Hardy D, Soeiro R, Fischl MA, Bartlett JG, Merigan TC, Hyslop NE, Richman DD, Valentine FT, Corey L and the AIDS Clinical Trials Group of the National Institute of Allergy and Infectious Diseases (1990)

Zidovudine in asymptomatic human immunodeficiency virus infection *New England Journal of Medicine* **322**: 941–949.
Volberding PA, Lagakos SW, Grimes J, Stein DS, Balfour HH Jr, Reichman RC, Bartlett JA, Hirsch MS, Phair JP, Mitsuyasu RT *et al.* (1994) The duration of zidovudine benefit in persons with asymptomatic HIV infection. Prolonged evaluation of protocol 019 of the AIDS Clinical Trials Group. *Journal of the American Medical Association* **272**(6): 437–442.
Whitehead J (1983) *The Design and Analysis of Sequential Clinical Trials*. New York: Halsted Press.
Whitehead J (1994) Sequential methods based on the boundaries approach for the clinical comparison of survival times. *Statistics in Medicine* **13**: 1357–1368.

9

Determining When a Data Monitoring Committee is Needed

Key Points

- All trials need careful monitoring, but not all trials need independent DMCs.
- Independent DMCs are most needed for randomized trials intended to provide definitive data regarding treatments intended to save lives or prevent serious disease.
- Independent DMCs are needed when interim analyses of safety and efficacy are considered essential to ensure the safety of trial participants.
- 'Internal' DMCs may be valuable in some trials that do not need independent monitoring

9.1 INTRODUCTION

All clinical trials require careful monitoring throughout their implementation. The primary reason for such monitoring is to identify any serious emerging safety concerns as rapidly as possible so as to minimize the time in which participants may be at excess risk. A second important reason is to identify problems in the conduct of the trial that could potentially be corrected, permitting successful completion of a trial that might otherwise not have met its objectives. Not all trials require a formal, independent DMC to achieve these ends, however. DMCs add a level of complexity to the conduct of a clinical trial, and also add costs for trial sponsors, so selectivity in their use is appropriate. In Chapter 1 we noted that DMCs have been used primarily in randomized trials of treatments intended to delay or prevent mortality or serious morbidity, and suggested some criteria to use in considering whether any particular trial ought to be monitored by an independent DMC. In this chapter we elaborate on these considerations and discuss in more detail particular settings in which DMCs could be useful. Other types of approaches to study monitoring may be considered when a fully independent DMC may not be necessary.

9.2 TYPICAL SETTINGS FOR AN INDEPENDENT DATA MONITORING COMMITTEE

Randomized trials testing treatments that may save lives or prevent progression of serious disease are by far the most common setting for the use of DMCs as they have been described in this book. In these trials, early results regarding safety and the primary efficacy endpoint could justify trial termination, even though secondary efficacy endpoints and long-term safety may not have been fully addressed. Further, the efficacy endpoint in such trials often has safety implications as well – a treatment intended to reduce mortality might turn out to have adverse effects that result in increased mortality, for example. Thus, such trials clearly require careful interim evaluations of comparative efficacy and safety data. Similarly, accumulating comparative data from trials performed in populations at high risk of mortality or major morbidity, even if the treatment is not directed at such endpoints (e.g., in individuals with advanced cancer, treatments for pain or agents to reduce the severity of chemotherapy-induced side-effects such as myelosuppression or cardiotoxicity), should also incorporate interim monitoring of comparative data if there is any possibility that the treatment could increase risk of death or other serious adverse outcomes that might be considered disease-related. Since judgments about interim data could potentially lead to modification or termination of the trial, they need to be made as objectively as possible; thus, independent DMCs should generally be established for such trials.

Before considering other settings in which an independent DMC might be needed, it is important to recognize the types of protection that are enhanced by a DMC in the traditional setting. In these trials, independent DMCs protect participants by ensuring that the trial is modified or terminated if there is persuasive evidence that participants are being put at unnecessary risk. This would be the case, for example, if the evidence became compelling before the planned completion of the trial that an investigational treatment is inferior to standard care. A DMC in this situation would recommend early termination so that no future patients would receive a clearly inferior treatment. Similarly (but less drastically), a DMC might recommend a modified dosing regimen or altered eligibility criteria if the frequency and/or severity of toxicity seen with the initial regimen were unacceptable.

DMCs are also needed to protect the welfare of future patients – not just future trial participants, but all patients with the disease or condition under study who will need treatment. As noted in Chapter 4, decisions made by individuals with a vested interest in the trial may be influenced by those interests, and such decisions may impact negatively on future patients. For example, if trial investigators were responsible for monitoring and interim decision-making, their concerns for patients they are treating could spur them to recommend termination on the basis of an early trend in favor of the control treatment, even when these early results might reasonably represent a chance finding that could easily be reversed with more follow-up. Or, even if they do not recommend termination, they might lose their enthusiasm for accruing patients, or for continuing to treat patients already enrolled as indicated by the protocol. Thus, a truly valuable new treatment might be lost to future patients unnecessarily. As another example, a trial sponsor for whom the trial outcome has major financial implications might be overly inclined to interpret early positive results as definitive. Early stopping in this case could be inadvisable, however, because non-definitive but suggestive early results are not likely to be found persuasive, either by a regulatory authority (in the case of an investigational therapy) or the medical community (in the case of a comparison of competing treatment strategies). Again, a valuable new treatment might be delayed in its application to patients who would benefit from it. Some sponsors, on the other hand, might be reluctant to stop early at all, even when results are definitive, for fear that the results will prove insufficiently persuasive. While it is difficult to predict how different vested interests might influence decision-making in particular situations, the independent DMC (as a body without significant vested interests) protects the integrity of the trial, as well as the trial participants. Protecting trial integrity usually has implications for the safety of future patients.

9.3 OTHER SETTINGS IN WHICH AN INDEPENDENT DATA MONITORING COMMITTEE MAY BE VALUABLE

9.3.1 Early trials of high-risk treatments

Some trials in other settings may also benefit from an independent DMC. For example, early phase trials of treatments with major risks might be in this category, even when not randomized. Generally, highly toxic interventions are used only to prevent major outcomes such as mortality or serious morbidity, since potentially severe toxicity would usually not be acceptable for treatments addressing lesser health outcomes; thus, the disease settings for such trials would be similar to those described in the previous section. In early phase trials of novel and potentially very toxic treatments, it might be desirable for independent DMCs to review emerging patterns of adverse outcomes, thereby reducing the likelihood that interim judgments would be influenced by financial or intellectual connections with the treatment under study. Institutional review boards and funding agencies have required on occasion that an independent DMC be established for early phase trials for novel treatments having the potential for significant adverse events. For example, independent DMCs have been established for some phase 1 and early phase 2 randomized trials evaluating the safety and biological activity of extracorporeal liver assist devices in the treatment of patients with acute liver failure, and for devices for management of patients in cardiac arrest (Mills *et al.*, 1999). DMCs have been advocated for gene therapy trials, even in the earliest phases, as a result of deaths observed in trials in which some felt that the study investigators who had primary responsibility for monitoring the trials did not give sufficiently serious consideration to initial adverse outcomes (Walters, 2000). Government-sponsored phase 1 trials of preventive HIV vaccines have been routinely randomized and placebo-controlled and have usually been monitored by an independent DMC (Ellenberg *et al.*, 1993). Finally, independent monitoring might be particularly valuable for early phase trials when financial or professional goals might be perceived to unduly influence the sponsor and/or investigators in the conduct of the trial, as, for example, when the sponsor is the individual who developed the product and is also testing it as the investigator.

9.3.2 Trials in vulnerable populations

Another type of trial for which an increased level of oversight might be warranted in some cases is where the study participants are considered vulnerable, with inadequate capability for protecting themselves (e.g., by refusing further treatment). In such trials, the potential participant cannot provide informed consent; rather, a relative or other legally authorized representative does so. Examples might be studies of antidepressant therapy in children, or of a drug to help maintain continence in individuals with Alzheimer's disease. In such trials, the therapies might not be expected to produce serious adverse effects, nor is the efficacy outcome one that would require early termination if efficacy were established earlier than anticipated. But an excess of even relatively minor adverse effects that would cause discomfort to participants (e.g., significant pruritus or nausea), and that was not clearly outweighed by the potential for improved quality of life should the product prove effective, should perhaps lead to changes in these studies; since participants in such trials may not be capable of protecting themselves by judging such effects to be intolerable and withdrawing from continued treatment on the study protocol, they may benefit from the extra protection afforded by an independent DMC.

9.3.3 Trials with potentially large public health impact

Finally, trials that are intended to have major public health impact, even if no major safety issues are anticipated, probably need an independent DMC. Trials evaluating and comparing currently available treatment strategies might be in this category. For example, while short-term trials of new antidepressants generally do not have (or need) DMCs, a long-term trial comparing multiple approaches to treating depression, with provision for initiating new approaches in those failing initial treatment, probably ought to have an independent DMC. A large and long-term clinical trial evaluating multiple strategies for treatment of depression, such as might be conducted by the National Institutes of Health, for example, might not raise major safety concerns but because participants would remain on their assigned treatment for a much longer period of time than would be the case in a typical trial evaluating a new antidepressant, the need for interim efficacy and safety analyses would be increased.

In such trials, each treatment studied is likely to have its own proponents in the medical community. Implementation of an independent DMC, consisting of scientists who are widely recognized for their expertise in the area being studied and who do not have financial or intellectual ties to any of the treatments under study, will improve the ability to safeguard the interests of trial participants and usually will enhance the credibility of trial results, particularly if the study is modified or terminated early. This could be very important for a trial that is intended to guide treatment choices for millions of patients.

It may be worth a reminder here that the existence of an independent DMC frees the trial leadership, who remain blinded to the interim comparative data, to implement changes to trial conduct and plans for analysis without risk of biasing the study.

9.4 AN ALTERNATIVE MONITORING APPROACH: THE 'INTERNAL' DATA MONITORING COMMITTEE

There are numerous clinical settings in which an independent DMC, with its complexities and costs, would not be essential and yet where it would still be desirable to have some of its oversight features – in particular, regularly scheduled meetings to review the accumulated safety experience, quality of trial conduct issues (e.g., completeness and timeliness of data capture, recruitment and eligibility rates, and levels of protocol adherence and patient retention), and newly available external information that could bear on the continuation of the trial as designed. For example, consider randomized trials that are conducted in settings with limited safety concerns but that are intended to provide primary evidence of product safety and efficacy. Randomized trials of interventions intended to provide short-term symptom relief, such as pain medications or asthma interventions might be in this category. For such trials, a fully independent committee usually would not be essential as long as there was no review of interim comparative outcome data (see Chapter 10), yet a sponsor might want some level of structured monitoring. In such cases, the study sponsor may establish an internal committee to regularly review the (blinded) interim data and formulate recommendations to trial leadership to help ensure optimal decision-making during the course of the trial.

Such oversight groups, which we will refer to as 'internal DMCs', perform many of the functions noted in Chapter 2. They will periodically monitor emerging data on efficacy and safety measures (usually only in the aggregate in controlled trials), as well as data on quality of trial conduct, to consider whether any changes in the conduct of the trial are warranted.

As with independent DMCs, internal DMCs should generally have multidisciplinary representation to achieve effective oversight of clinical, statistical and operational aspects of the trial. In industry trials, some sponsors have constituted internal DMCs with a group of individuals from a variety of disciplines who are not involved in the trial's design and conduct – for example, representatives of the sponsor's senior-level leadership in the areas of clinical science and regulatory affairs, as well as other company clinicians and biostatisticians who have no operational responsibilities for the trial (Stump, personal communication, 2000; Hopkins, personal communication, 2000). This membership could be entirely internal to the industry sponsor or may be supplemented by one or two experts external to the company. An internal committee constituted in this way provides some level of independent oversight, and might identify some concerns or data patterns that the study leadership group, focused on the day-to-day conduct, might miss. While such review might not be considered adequately independent for those studies that do require interim review of comparative data, for those studies that do not an internal DMC could provide advice about safety and/or trial conduct issues to the trial leadership during the course of the trial that might be useful in many settings.

Other important settings in which internal DMCs could be engaged are early phase trials with major concerns about safety but for which independent DMCs have not been implemented. In oncology research, for example where cancer centers and cooperative groups funded by the National Cancer Institute are required by the NCI to have interim monitoring programs (National Cancer Institute, 2001) for all clinical trials, internal DMCs have often been used for monitoring phase 1 and early phase 2 trials. For these early phase non-randomized trials, the internal DMC has frequently included the trial leadership. The Southwest Oncology Group (SWOG), for example, has for many years relied on internal DMCs to monitor its phase 1 and early phase 2 trials designed to obtain early data on safety and tumor response measures, to select doses for further studies, and to address safety and biological activity. SWOG's clinical and statistical leadership have served on these internal DMCs to provide oversight of trial conduct issues and to safeguard patient interests.

In some settings it might be preferable for an internal DMC to have sole access to certain interim data, even when the study is uncontrolled or when only aggregate data are evaluated during the trial. As noted in Chapter 5, aggregate event rates in a controlled trial may sometimes suggest an emerging treatment effect, or lack of one; in an uncontrolled study, an interim observed response or 'success' rate might influence investigators' willingness to continue to enter patients, even when relatively few patients have been treated and these rates are unstable. For example, in the early phase trials conducted by SWOG, the internal DMCs have improved certain aspects of the trials program by maintaining confidentiality of the accumulating data. Prejudgment of early results from these trials – with tapering off of accrual if no responses were observed in the first few patients – has been prevented by SWOG through its policy that interim data would only be provided to the internal DMC.

Committees given sole access to certain emerging data would need to report back to the trial leadership, just as independent DMCs need to provide feedback to those conducting and managing the trial. There is no reason why an internal DMC could not follow the same basic model as described in Chapter 6 for an independent DMC, meeting regularly with the trial's clinical and statistical leadership in open sessions, and discussing interim results (considering aggregate data only) and developing recommendations to the trial leadership in closed sessions attended only by committee members and those performing the analyses. Minutes of each session would provide a record of the proceedings and a summary of any recommendations made.

Internal DMCs may also be formed to monitor all trials of a new product under development, rather than one single trial. A mandate for the monitoring of an entire development program, rather than a single trial, could be particularly valuable in that a committee reviewing data from multiple trials of the same product might notice rare events or patterns of responses that would not be discernible in a data set for a single study.

9.5 A DECISION MODEL FOR ASSESSING NEED FOR AN INDEPENDENT OR INTERNAL DATA MONITORING COMMITTEE

Table 9.1 sets out considerations for addressing the need for a structured monitoring approach – either an independent or an internal DMC – in a clinical trial. We categorize trials into one of two settings, represented in the tiers of the table. Setting 1 consists of trials with higher levels of concerns about safety. These include trials of interventions to prevent or treat diseases that lead to death or irreversible morbidity, trials of novel treatments with potential to induce significant and perhaps unpredictable adverse events, and trials conducted in vulnerable populations such as children, the elderly, and incarcerated or mentally compromised patients. Setting 2 includes all other trials, and is by far the larger category.

The table's columns show how key factors lead to a decision about the approach to interim monitoring. Two factors are considered: the potential level of ethical concerns, and concerns about trial integrity and credibility of results. Ethical concerns relate to the potential for harm to study participants – harm that would be incurred, for example, if an inferior treatment were to continue to be given beyond the point at which its inferiority had been definitively established. Integrity/credibility concerns relate more to the potential impact of the trial on public health – for example, its importance to decision-making by regulatory bodies which will determine whether a new treatment will be made available to the public. These two factors clearly are interrelated, but both need to be

Table 9.1 Settings for an independent DMC or an internal DMC

Type of setting[1]	Imperatives		Need for DMC	
	Ethical	Credibility/ integrity	Independent DMC	Internal DMC
Setting 1				
Randomized trials (Ph 2b, 3, 4)	YES	YES	YES	–
Randomized trials (Ph 1, 2a)	YES	likely	maybe	likely[2]
Non-randomized trials	YES	maybe	unlikely	likely[2]
Setting 2				
Randomized (any phase trial)	unlikely	likely	unlikely[3]	maybe[2]
Non-randomized	unlikely	unlikely	NO	unlikely

[1] Setting 1 includes: life-threatening diseases (treatment, palliation and prevention); diseases causing irreversible serious morbidity (treatment, palliation and prevention); novel treatments for life-threatening diseases (treatment, palliation and prevention) with potential for significant adverse events; vulnerable populations. Setting 2 includes trials not included in setting 1.
[2] An internal DMC would be advised if an independent DMC were not established.
[3] Integrity/credibility issues could motivate use of an independent DMC; for example, if a trial in this setting were to impose interim monitoring of comparative data.

considered. If we wish to stop a trial before its planned completion because one of the study treatments has been conclusively established to be inferior to the other, a group of experts needs to be looking at the interim data at regular intervals. In order to address ethical concerns and to preserve study integrity – that is, to minimize the chance that trial results will be biased – this group of experts should be independent of the study leadership, as discussed in Chapter 4.

As we have noted, independent DMCs are most commonly used for phase 3 randomized trials conducted in setting 1, but may be desirable on occasion for earlier phase trials, particularly when concerns about the safety of participants or the conflicts of interest in monitoring the trial would be unusually strong. For those trials in setting 1 where an independent DMC will not be implemented, establishing an internal DMC would be advised.

Trials of any phase performed in setting 2, in which difficult decisions regarding safety of trial participants are less likely to arise, do not generally require the involvement of a formal, independent DMC. When the primary outcome variable of a trial is of a lesser order than mortality or an event with major impact on health status such as myocardial infarction, stroke or tumor recurrence, there may be less of an ethical concern about carrying a study to its planned completion even if efficacy is so great that it is clearly established part-way through the trial. One might even argue that, for these less serious outcomes the ethical considerations move one in the direction of completing the trial even in the face of established efficacy because, for products with a less than major health benefit, it is very important to develop a full safety profile before the product becomes widely available. If there is no compelling ethical reason to stop early because of efficacy results, as is typically the case for such trials, the need to review the efficacy during the course of the trial is limited. In fact, trials of low-risk treatments for relatively minor conditions are usually performed with careful safety monitoring (considering aggregate data only) but no interim efficacy analyses.

Further, occurrence of serious toxicity is unusual in such trials; individual cases are not difficult to identify and are carefully reviewed by sponsors and investigators as well as by regulatory reviewers, to whom such cases must be reported rapidly. For example, in a short-term trial of a new antihistamine, in which participants would be treated on study for only 2–4 weeks, the types of toxicity that would be expected at low rates (and could be tolerated if the product proved effective) might include nausea, constipation or diarrhea, headache, or other mild symptoms. A death or an episode of major morbidity (myocardial infarction, stroke, liver or kidney failure, etc.) would be immediately reported to the regulatory authority and to all participating investigators and their institutional review boards, quite possibly resulting in the (at least) temporary suspension of the trial. The event would be thoroughly investigated by the sponsor and the regulatory authority; independent judgment would not be required regarding the need to review the case carefully or to evaluate the appropriateness of continuating the trial. Evaluation of such events would generally not depend on a comparison of rates of the event in the treatment and control arm, and any level of serious toxicity detectable in clinical trials would generally be considered unacceptable for a product that did not prevent comparably serious health events. For randomized trials in these settings, interim comparisons of efficacy and safety data are therefore not typically needed; such data generally remain blinded throughout the course of the trial, as there would be no compelling reason to terminate the trial early based on differences in efficacy outcomes or in the occurrence of minor adverse events.

Even in setting 2, however, an independent DMC might be valuable in some circumstances. If a trial may be expected to have an important public health impact, even if it does not meet the criteria for setting 1, the use of an independent monitoring group may be warranted because of the need for a very high level of study credibility, as discussed earlier.

A potential reason for engaging an independent DMC rather then an internal DMC in a setting 2 trial would be the need to perform interim comparative outcome analyses during the trial. Although randomized trials in setting 2 usually are performed without review of interim treatment comparisons, as we have discussed, in some cases such analyses could be desirable. For example, if the trial involved a fairly long course of treatment and/or if accrual was expected to continue over an extended period, it might be desirable, both for ethical and economic reasons, to evaluate the data part-way through the trial. Interim analyses, possibly leading to early termination, could reduce the extent to which trial participants would be exposed to an ineffective treatment, and would preserve resources for more promising endeavors.

9.6 SETTINGS WITH LITTLE NEED FOR AN INDEPENDENT OR INTERNAL DATA MONITORING COMMITTEE

Some clinical trials do not require either an independent or an internal DMC. An illustration would be early phase non-randomized trials in settings with limited safety concerns. In such settings, the data will be considered preliminary; while bias in such trials is not completely innocuous (improper conclusions in early phase trials can lead to suboptimal study designs in later phase trials and thus reduced chance of identifying the optimal use of the treatment), it is unlikely that one would need a formal DMC structure of any type to adequately monitor for safety and trial conduct concerns. Even here, however, a committee overseeing the entire development program (rather than individual trials) could have potential utility.

In some clinical trials having very rapid recruitment and short-term endpoints, there may be no practical way for a DMC to provide meaningful oversight. For example, suppose enrollment can be completed in less than six months and the principal assessments of safety and efficacy endpoints are planned to occur over the first 30 days after initiation of treatment. A reliable and substantive presentation of interim data before the end of enrollment and study treatment might not be possible. If interim monitoring is imperative in such settings, one would need to conduct the trial so that either enrollment was slowed, or enrollment was suspended for some duration after some proportion of the desired sample size had been enrolled. Alternatively, extraordinary data management procedures might be implemented to allow real-time monitoring of relatively current and accurate data.

9.7 SUMMARY

As noted in Chapter 1, independent DMCs have been most frequently implemented in randomized clinical trials designed to address the benefit-to-risk profile of an intervention, or the relative profiles of two or more regimens, in a setting that addresses major health outcomes such as mortality, progression of a serious disease, or occurrence of a life-threatening event such as heart attack or stroke.

In most early phase and non-randomized trials in setting 1 and in many trials in setting 2, as defined in Table 9.1, an internal DMC may provide some of the features of oversight by an independent DMC, while reducing some of the complexities and cost. Such an approach could be desirable for monitoring randomized trials (nearly always short-term) that address symptom relief, or trials implemented early in drug development whose results will be examined in an exploratory fashion.

REFERENCES

Ellenberg SS, Myers MW, Blackwelder WC, Hoth DF (1993) The use of external monitoring committees in clinical trials of the National Institute of Allergy and Infectious Diseases. *Statistics in Medicine* **12**: 461–467.

Mills JM, Maguire P, Cronin DC, Conjeevarum H, Johnson R, Conlin C, Brotherton J, O'Laughlin R, Triglia D, Piazza R (1999) ELAD continuous liver support system: Report of cell function (abstract). *Hepatology* **30**: 168A.

National Cancer Institute (2001) *Essential Elements of a Data and Safety Monitoring Plan for Clinical Trials funded by the National Cancer Institute*, April 4. http://cancertrials.nci.nih.gov/researchers/dsm/dsm.pdf.

Walters L (2000) *Statement before the Subcommittee on Public Health, Senate Health and Education Committee*, February 2. http://www.senate.gov/ labor/hearings/feb00hrg/020200wt/frist0202/gelsing/kast/patter/fda-zoon/verma/walters/walters.htm.

10

Regulatory Considerations for the Operation of Data Monitoring Committees

Key Points

- FDA regulations require ongoing safety monitoring of clinical trials, but DMCs are only minimally addressed in government regulations.
- Despite lack of requirements for DMCs, regulators generally expect randomized trials with mortality or major morbidity endpoints to be monitored by an independent DMC.
- FDA reviewers generally do not participate in DMC meetings or serve on DMCs of trials they regulate, and there are good reasons why they should not.
- The FDA issued a draft guidance document on the establishment and operation of DMCs in 2001.

10.1 INTRODUCTION

The establishment and operation of formal DMCs have been addressed only minimally by regulatory authorities in the USA and elsewhere. The absence of attention to this critical aspect of clinical trial conduct within the large body of regulatory documents pertaining to clinical trials of investigational products is almost surely due to the fact that, until the 1990s, trials sponsored by the pharmaceutical industry rarely made use of such committees. This situation has been rapidly changing, and commentary on the use of DMCs began to enter the regulatory literature in the late 1990s. A Food and Drug Administration draft guidance document on this topic has recently been issued.

10.2 DATA MONITORING COMMITTEES IN FDA REGULATIONS AND GUIDANCE

The first and only mention of DMCs in the US Code of Federal Regulations appeared in 1996, in a new regulation addressing requirements for carrying out studies in emergency circumstances in which the obtaining of informed consent from the individual to be treated or a family member of that individual is not feasible (Code of Federal Regulations, Title 21, Part 50.24). This regulation was implemented because certain studies of promising new treatments for victims of trauma and sudden cardiac arrest, who were generally unconscious and for whom identification and tracking down of relatives to obtain informed consent were unlikely to be accomplished in a short time, appeared to be totally prohibited by then-existing regulations of the FDA and of the Department of Health and Human Services (DHHS) with waiver of informed consent. The development of the new regulation, providing for the conduct of such studies in limited circumstances, included an extensive set of protections beyond those normally required for study of investigational drugs, biologics and medical devices. One protection was the requirement for 'Establishment of an independent data monitoring committee to exercise oversight of the clinical investigation . . .'. The preamble to this regulation noted that a variety of models for the operation of such committees are in use, and referred to the publication of conference proceedings in which such models are discussed and critiqued (Ellenberg *et al.*, 1993), but did not provide specific guidance or direction. The monitoring committee experience of one of the first trials conducted under this regulation has been described by Lewis *et al.* (2001).

Data monitoring committees have been briefly addressed in guidance documents issued by the FDA. Probably the earliest of these, issued in 1988, is entitled *Guideline for the Format and Content of the Clinical and Statistical Sections of New Drug Applications* (US FDA, 1988), often referred to informally as the 'Clinstat Guideline'. This document includes a single paragraph on interim data monitoring, with a mention of DMCs:

> The process of examining and analyzing data accumulating in a clinical trial, either formally or informally, can introduce bias. Therefore, all interim analyses, formal or informal, by any study participant, sponsor staff member, or data monitoring group should be described in full, even if the treatment groups were not identified. The need for statistical adjustment because of such analyses should be addressed. Minutes of meetings of a data monitoring group may be useful (and may be requested by the review division).

Guidance documents developed through the International Conference on Harmonisation (ICH), a collaboration of industry and regulatory authorities in the USA, Europe and Japan to establish consistent regulatory requirements worldwide, also refer to independent DMCs. These documents, all in the ICH 'efficacy' series (other series focus on aspects of quality and safety), include E3: Structure and Content of Clinical Study Reports (ICH, 1995) E6: Good Clinical Practice:

Consolidated Guideline (ICH, 1996) and E9: Statistical Principles for Clinical Trials (ICH, 1998). E3 indicates that clinical study reports submitted to the regulatory authority should contain information about the composition and operating procedures of any DMC involved with the trial, and that the interim data reports and minutes of DMC meeting should be included as an appendix to the clinical study report. E6 affirms that good clinical practice includes maintenance of written procedures and written records for DMC activities. E3 and E6 do not go into any details regarding DMC operations, however.

E9 does begin to give a little guidance regarding the establishment and operation of these committees, in addition to providing a substantial discussion of the statistical issues relating to interim analysis of study data. With regard to the establishment of DMCs, the document states:

> For many clinical trials of investigational products, especially those that have major public health significance, the responsibility for monitoring comparisons of efficacy and/or safety outcomes should be assigned to an external independent group, often called an Independent Data Monitoring Committee (IDMC), a Data and Safety Monitoring Board or a Data Monitoring Committee whose responsibilities need to be clearly described. (ICH, 1998, p. 20)

A few paragraphs later, the following brief section appears, describing the role of an IDMC:

> An IDMC may be established by the sponsor to assess at intervals the progress of a clinical trial, safety data, and critical efficacy variables and recommend to the sponsor whether to continue, modify or terminate a trial. The IDMC should have written operating procedures and maintain records of all its meetings, including interim results; these should be available for review when the trial is complete. The independence of the IDMC is intended to control the sharing of important comparative information and to protect the integrity of the clinical trial from adverse impact resulting from access to trial information. The IDMC is a separate entity from an Institutional Review Board (IRB) or an Independent Ethics Committee (IEC), and its composition should include clinical trial scientists knowledgeable in the appropriate disciplines including statistics.
>
> When there are sponsor representatives on the IDMC, their role should be clearly defined in the operating procedures of the committee (for example, covering whether or not they can vote on key issues). Since these sponsor staff would have access to unblinded information, the procedures should also address the control of dissemination of interim trial results within the sponsor organisation. (ICH, 1998, p. 21)

These documents, while noting the use and operation of DMCs, provide little in the way of guidance as to the appropriate composition and function of such committees. As this book was being completed, the FDA issued a draft guidance specifically focused on DMCs, which will be discussed in section 10.6.

The minimal reference to DMCs in formal regulation and guidance does not imply that regulators have been indifferent to the use of DMCs. FDA review divisions generally expect to see DMCs established for randomized trials with mortality or major morbidity as primary endpoints. Further, it is not unusual for FDA reviewers to consider and comment on the data monitoring approaches specified in protocols submitted for prior review. It is both natural and appropriate for product regulators to take a major interest in the plans for data monitoring, and the structure and operations of the DMC that will carry out this monitoring, since the credibility and value of the study results, as well as the welfare of study participants, may depend critically on scientifically sound approaches to reviewing (and possibly acting on) the interim analysis of trial data. It is therefore important that the proposed approach to data monitoring and operation of the DMC be clearly laid out in the study protocol that is submitted to the regulatory agency before the initiation of the study (ICH, 1998; O'Neill, 1993).

10.3 REGULATORY REQUIREMENTS RELEVANT TO DATA MONITORING COMMITTEE OPERATION

While FDA regulations do not deal specifically with DMCs (except in the specific case of studies allowing waiver of informed consent, as discussed earlier), they do address the need for regular monitoring of the study with regard to the safety of current and future participants. Study sponsors are required to maintain assurance as the study progresses that the product under investigation remains safe, and to terminate immediately the study of any product found to be associated with unreasonable and significant risk. Reports of serious and unexpected adverse reactions to the investigational product must be reported to the FDA within a short time following their occurrence; annual reports in which the ongoing study experience is summarized (with particular emphasis on the adverse experience reports) are also required. The regulations do not specify that interim efficacy data must be assessed by the sponsor, but do require the submission of any interim efficacy results known to the sponsor at the time of the annual report. These regulations, requiring rapid reporting of certain adverse reactions and regular summaries of interim results known to the sponsor, implicitly establish the joint responsibility of the sponsor and the FDA for ensuring that ongoing studies continue to be safe and appropriate. FDA involvement is emphasized even more strongly in the section describing special procedures for expedited investigation of products intended to treat life-threatening illnesses: 'For drugs covered under this section, the Commissioner and other agency officials will monitor the progress of the conduct and evaluation of clinical trials and be involved in facilitating their appropriate progress' (Code of Federal Regulations, Title 21, Part 312.87). This section of the regulations, motivated primarily by the desire to make promising AIDS treatments available as rapidly as possible, demonstrated the intent of the FDA to be active in addressing important public health concerns. While there is no explicit reference to accessing interim data, the wording above may suggest a potential conflict with common DMC policies against sharing the results of interim data analyses with regulatory authorities. Thus it is understandable that as the use of DMCs has increased, particularly in the setting of new products for serious diseases, there has been occasional uncertainty about the impact of an independent DMC on the appropriate role of FDA staff in the oversight process.

In some types of study, the number of serious adverse events is large and there may be substantial overlap between known adverse events and events associated with progression of disease. For example, in large multicenter studies of new treatments for individuals with a high risk of mortality, it may not be clear whether any particular patient death is attributable to the drug or to the disease the drug is treating. In such studies, review of the individual adverse event reports is not useful in assessing the interim safety experience, particularly since these reports are generally reviewed without treatment codes by the regulatory reviewer as well as the pharmaceutical company. In these circumstances, regular review of tabular summaries of adverse events by treatment arm is critical to the ongoing assurance that study participants are not at undue risk. Such reviews are optimally performed by a DMC since, as discussed earlier, others associated with the trial (including the regulatory reviewers) will generally remain blinded to the comparative interim data during the course of the trial.

10.4 INVOLVEMENT OF FDA STAFF IN DATA MONITORING COMMITTEE DELIBERATIONS

There is no written guidance for FDA reviewers regarding their involvement in the formal data monitoring process. FDA reviewers always have some degree of involvement in monitoring the process of study investigations, as noted above; they review safety data from ongoing studies that are submitted to the Agency on time schedules dependent on the severity of the events being reported, as well as the annual reports on study progress (Code of Federal Regulations, Title 21, Parts 312.32–33). Nevertheless, an informal policy of keeping FDA reviewers separate from DMC deliberations has developed. As O'Neill (1993) states, there is 'a general consensus . . . that the FDA should not and does not want to be a routine observer nor a voting member in a DSMB'. The rationale is straightforward: an FDA reviewer who participates in a DMC decision to terminate a study early due to apparently strong efficacy may be somewhat compromised in his/her ability to carry out a neutral and objective regulatory review; such a reviewer might find it more difficult to advocate any action other than approval, even if information available to him/her as an FDA reviewer gives a different picture than the more limited data available to the DMC at the time of decision-making. In addition, FDA must review and approve interim changes in the protocol; if FDA reviewers were aware of the interim data, they would be hampered in making objective assessments of proposed changes. For example, if a sponsor proposed, based on new data from other studies, to change the primary endpoint of a trial, it could be difficult for the FDA reviewer to consider this in a purely objective way if the reviewer knew whether the change would improve or reduce the chance that the study would ultimately demonstrate benefit of the product. In general, therefore,

FDA staff do not participate as members or observers in DMCs for studies of products they review.

An early experience with FDA staff accessing interim data demonstrates the type of problem that could arise. An FDA medical reviewer was provided online access to accumulating data for an ongoing trial of a product under his review. As the data began to accumulate, the reviewer became increasingly concerned about safety considerations. Ultimately he argued that the trial should be terminated for safety reasons, although only a small number of deaths had been observed. Later on, after the data had been subject to a more thorough review, there was consensus that the trial results were not definitive and the trial probably should not have been terminated. It is possible that the FDA reviewer overreacted to the emerging data because of concern that he might be held accountable for permitting a trial of an unsafe product to continue for too long. This experience contributed to the informal policy noted above, that FDA staff would generally refrain from routine involvement with interim review of ongoing trials (Temple, personal communication, 2001; Task Force of the Working Group on Arrhythmias of the European Society of Cardiology, 1994).

FDA staff have occasionally served on DMCs without a 'regulatory hat'; that is, they are not involved with the regulatory review of the product being investigated. This situation would seem to pose fewer problems than the situation described above, although there could still be concern that an FDA reviewer's interpretation of the results of a particular trial might be affected by knowing that an FDA colleague served on the DMC for that trial. Also, it would be important for the trial sponsor and other DMC members to understand that the opinions of the FDA member could not be viewed as predictive of the ultimate regulatory assessment of the data.

As the use of DMCs increases in regulated trials, it will be important for FDA reviewers to be familiar with typical DMC objectives and operations, in the same way that they need to be familiar with other aspects of clinical trials. For this reason, the occasional participation of FDA reviewers on DMCs of products for which they have no regulatory responsibility may have important educational value, in addition to the value an experienced FDA reviewer might bring to the monitoring process.

10.5 EXAMPLES OF FDA INTERACTION WITH DATA MONITORING COMMITTEES

While the FDA has in practice accepted the concept of DMCs as the sole group reviewing the (unblinded) interim efficacy data of some clinical trials, there is substantial rationale and precedent for the FDA to take a somewhat more active role in certain circumstances. An FDA reviewer, for example, might raise concerns if a proposed DMC for a study under review appeared inappropriate – if no statistician or no clinician with relevant specialized expertise were included, for example, or if any of the members appeared to have a significant conflict of interest. Such concerns would have to be resolved through discussion and negotiation, since there are currently no regulations specifying anything about DMC membership.

In NIH-sponsored studies of treatments for HIV infection and its sequelae, FDA staff have had significant opportunity for input into the monitoring process even though they have not participated in the review of interim data comparisons. As described in Chapter 7, FDA representatives have had regular opportunities to meet DMC members and others involved with the trial to discuss trial progress, safety issues, and other issues outside the review of unblinded interim data. This model has worked well in an area of high visibility, where the need for urgency in making products available as soon as possible after they have been demonstrated to be effective and safe, has led the FDA to take unusual steps to expedite product reviews.

One particular FDA–DMC interaction demonstrates the type of productive joint effort that can be made when circumstances demand unusual action. In 1990, clinical trials comparing the then-investigational antiviral therapy ddI to AZT, the only antiviral therapy then available for treatment of HIV infection, were ongoing under NIH sponsorship. By this time, many individuals who had initially done well on AZT had become intolerant to it or were declining in spite of it; there was great anticipation that the new therapy might avert death of large numbers of AIDS patients. The FDA wanted to review efficacy data as rapidly as possible so as not to delay the availability of this new therapy should it appear effective. New regulations were under development to permit accelerated approval of potentially life-saving new therapies on the basis of 'surrogate endpoints' – laboratory tests or other markers that were thought to predict serious clinical outcomes such as death or disease progression. The FDA was prepared to consider approving ddI on this basis, even though the regulation was not yet finalized. The manufacturer of ddI had some promising phase 1 data on CD4+ cell counts following administration of ddI, but the FDA felt the need for far more data to justify an approval, given that this approach was such a major departure from the usual process of evaluating and approving new drugs. The Director of the Center for Drug Evaluation and Research arranged to meet with the DMC overseeing the NIH-sponsored ddI studies to discuss the possibility of making interim marker data available from the ongoing studies.

The DMC was initially very reluctant to consider such a proposal. Many individuals at that time believed strongly that change in the CD4+ cell count was an accurate measure of drug effect on clinical outcomes (a belief that was later called into question by further studies (Choi *et al.*, 1993; Concorde Coordinating Committee, 1994; DeGruttola *et al.*, 1993)). The DMC was concerned that release of these data while the study was ongoing might compromise the ability to complete the study successfully and obtain a reliable assessment of the clinical effect of ddI treatment. But the DMC members were appreciative of the public health issue, and an arrangement was ultimately made to provide marker data from one of the three ongoing ddI studies. The study selected was only a few months short of its projected completion date, and it was clear that the effect on the treatment estimate of even a large number of study participants switching their treatment as a result of the public reporting of the interim marker data would be minimal. Thus, the FDA was able to meet its objective of making promising new therapies available rapidly, and the DMC was able to meet its responsibility to ensure the integrity of the studies it monitored so that a reliable answer to the clinical question could ultimately be obtained. This satisfactory outcome was possible because all participants in the negotiations appreciated the legitimacy of both the FDA's and the DMC's goals in this complex situation and were willing to modify their preferred approaches in order to achieve these goals.

There are examples of more recent interactions. In one case, a manufacturer submitted a licensing application to the FDA with the primary evidence of efficacy being the result of a large, well-conducted placebo-controlled randomized trial. A second trial, similar in design, was under way, but the company chose to submit a licensing application using only the data from the first trial, based on the view that the new treatment represented an important advance and it would be inappropriate to delay seeking licensure while awaiting the results of the second trial. The FDA was uncomfortable about licensing the product knowing that highly relevant additional data had been collected but were not yet available. In particular, the FDA was aware that the DMC for the second trial had reviewed substantial interim data and, despite being aware of the positive results of the first trial, had recommended that the trial continue. On the other hand, FDA staff were concerned, for reasons discussed earlier and in Chapter 7, about requesting access to interim data and using such data to inform decision-making. Knowledge of interim data could prevent FDA reviewers from making fully objective judgments regarding proposals for interim changes in trial design. Further, if the FDA were known to routinely (or even frequently) review interim data from ongoing trials, clinical trials could be undermined; those with interests in trial results, ranging from patient advocates to investment analysts, would broadcast their speculations on trial trends for the presumed benefit of their constituents. In this particular case, however, the FDA office regulating the product in question did decide to approach the sponsor to request access to the interim data and an opportunity to meet with the DMC of the ongoing study. The sponsor agreed to this request. At the meeting the interim data were discussed and compared with the data reported for the earlier trial. Discussion also focused on differences in the design and conduct of the two trials, the potential impact of the data from the first trial on the ability to successfully complete the second trial, and the likelihood that continuation of the second trial would provide significant additional information on the effects of the drug under investigation. Because the interim data reviewed were inconsistent with the final results of the original trial, regulatory action was delayed. Ultimately, the second trial was completed with final results still not supportive of the results of the original trial, leaving open the question of the efficacy of the product for the clinical condition under study.

It must be emphasized that such examples are, and should be, rare, because of the concerns raised earlier in this chapter.

10.6 FDA DRAFT GUIDANCE ON DATA MONITORING COMMITTEES

In 1998, the Office of the Inspector General of the DHHS issued a call for reform of procedures to ensure the safety of patients participating in clinical trials (DHHS, 1998). The bulk of the recommendations addressed the functioning of institutional review boards, but some specific recommendations were made in regard to DMCs. In particular, the FDA was urged to provide more guidance on such issues as what sorts of trials should have DMCs, how these committees should operate and how their responsibilities could be more effectively integrated with those of IRBs to provide higher assurances for the safety of trial participants. Government officials agreed with these recommendations and moved to implement them (Shalala, 2000).

In 2001, a draft guidance document entitled *Guidance for Clinical Trial Sponsors on the Establishment and Operation of Clinical Trial Data Monitoring Committees* was issued (US FDA, 2001). As a guidance document, it does not impose any requirements on sponsors of clinical trials; rather, it describes possible approaches that the FDA would deem acceptable. We will discuss this document only briefly, as it is a draft and will be subject to revision following receipt of public comment.

The document addresses many of the issues discussed throughout this book. It provides some general information on DMC structure and operation that may be particularly useful to that subset of trial sponsors with limited experience in working with DMCs, and it discusses the meaning and importance of DMC independence from the trial sponsor. It includes considerations for selecting committee members, structuring committee meetings, specific monitoring responsibilities that might be assigned to a DMC, and determining whether a given study would benefit from oversight by a DMC. It also discusses the relative roles of a DMC and the study sponsor in meeting the safety monitoring requirements laid out in FDA regulations.

A particular concern highlighted in the document is the need to protect the integrity of the study from possible influence by interim results. Sponsors are advised to establish a study structure that isolates those with knowledge of interim results (i.e., those individuals involved in preparing interim analyses for review by a DMC) from those with responsibility for managing the study, to the extent possible. Sponsors are also alerted to the potential difficulties that the FDA could face in interpreting final study results when interim results had been available to the sponsor. In such cases, the possibility that knowledge of the interim data influenced further conduct of the study can never be completely excluded.

After a period of public comment, the FDA will revise the document to incorporate appropriate modifications, additions and deletions suggested by the public to improve its quality, clarity and usefulness.

REFERENCES

Choi S, Lagakos SW, Schooley RT, Volberding PA (1993) CD4+ lymphocytes are an incomplete surrogate marker for clinical progression in persons with asymptotic HIV infection taking zidovudine. *Annals of Internal Medicine* **118**: 674–680.

Concorde Coordinating Committee (1994) Concorde: MRC/ANRS randomised double-blind controlled trial of immediate and deferred zidovudine in symptom-free HIV infection. *Lancet* **343**: 871–881.

DeGruttola V, Wulfsohn M, Fischl M, Tsiatis A (1993) Modeling the relationship between survival and CD4+ lymphocytes in patients with AIDS and AIDS-related complex. *Journal of the Acquired Immune Deficiency Syndrome* **6**: 359–365.

Department of Health and Human Services (1998) *Institutional Review Boards: Their Role in Reviewing Approved Research* (OEI-01-97-00190). Office of Inspector General, June.

Ellenberg SS, Geller N, Simon R, Yusuf S (eds) (1993) Proceedings of 'Practical issues in data monitoring of clinical trials', Bethesda, Maryland, USA, 27–28 January 1992. *Statistics in Medicine* **12**: 415–616.

International Conference on Harmonisation (1995) *Structure and Content of Clinical Study Reports*, Guideline E3. http://www.ifpma.org/ich5e.html.

International Conference on Harmonisation (1996) *Guideline for Good Clinical Practice, Guideline E6*. http://www.ifpma.org/ich5e.html.

International Conference on Harmonisation (1998) *Statistical Principles for Clinical Trials*, Guideline E9. http://www.ifpma.org/ich5e.html.

Lewis RJ, Berry DA, Cryer III H, Fost N, Krome R, Washington GR, Houghton J, Blue JW, Bechofer R, Cook T, Fisher M (2001) Monitoring a clinical trial conducted under the Food and Drug Administration regulations allowing a waiver of prospective informed consent: the Diasporin Cross-Linked Hemoglobin Traumatic Hemorrhagic Shock Efficacy Trial. *Annals of Emergency Medicine* **38**: 397–404.

O'Neill RT (1993) Some FDA perspectives on data monitoring in clinical trials in drug development. *Statistics in Medicine* **12**: 601–608.

Shalala D (2000) Protecting research subjects – what must be done. *New England Journal of Medicine* **343**: 808–810.

Task Force of the Working Group on Arrhythmias of the European Society of Cardiology (1994) The early termination of clinical trials: Causes, consequences and control with special reference to trials in the field of arrhythmias and sudden death. *Circulation* **89**: 2892–2907.

US Food and Drug Administration (1988) *Guideline for the Format and Content of the Clinical and Statistical Sections of an Application*. Rockville, MD: FDA. http://www.fda.gov/cder/guidance/statnda.pdf.

US Food and Drug and Administration (2001) *Guidance for Clinical Trial Sponsors on the Establishment and Operation of Clinical Trial Data Monitoring Committees*. Rockville, MD: FDA. http://www.fda.gov/cber/gdlns/clindatmon.htm.

Appendix A

The Data Monitoring Committee Charter

TITLE OF PROTOCOL:

PROTOCOL NUMBER:

SPONSOR OF PROTOCOL:

DATE OF DOCUMENT:

TABLE OF CONTENTS

1. INTRODUCTION

This Charter is for the Data Monitoring Committee (DMC) for [***provide protocol number***], for [***provide protocol title***].

The Charter will define the primary responsibilities of the DMC, its relationship with other trial components, its membership, and the purpose and timing of its meetings. The Charter will also provide the procedures for ensuring confidentiality and proper communication, the statistical monitoring guidelines to be implemented by the DMC, and an outline of the content of the Open and Closed Reports that will be provided to the DMC.

Definition of terms and abbreviations to be used in the Charter: [***fill in this information***].

2. PRIMARY RESPONSIBILITIES OF THE DMC

The DMC will be responsible for safeguarding the interests of trial participants, assessing the safety and efficacy of the interventions during the trial, and for monitoring the overall conduct of the clinical trial. The DMC will provide recommendations about stopping or continuing the trial. To contribute to enhancing the integrity of the trial, the DMC may also formulate recommendations relating to the selection/recruitment/retention of participants, their management, improving adherence to protocol-specified regimens and retention of participants, and the procedures for data management and quality control.

The DMC will be advisory to the clinical trial leadership group, hereafter referred to as the Steering Committee (SC) and usually including a sponsor representative. The SC will be responsible for promptly reviewing the DMC recommendations, to decide whether to continue or terminate the trial, and to determine whether amendments to the protocol or changes in study conduct are required.

3. ORGANIZATIONAL DIAGRAM

The following diagram shows the relationships between the DMC and other committees and functional areas involved in the trial.

[***The diagram providing the Organizational Diagram should be inserted here.***]

4. MEMBERSHIP OF THE DMC

4.1 Members

The DMC is an independent multidisciplinary group consisting of biostatisticians and clinicians that, collectively, has experience in the management of patients with [***fill in disease***] and in the conduct and monitoring of randomized clinical trials.

DMC Chair:	***Address*** ***Telephone/Fax*** ***Email address***
DMC Biostatistician:	***Address*** ***Telephone/Fax*** ***Email address***
DMC Clinical Investigators:	***Address*** ***Telephone/Fax*** ***Email address***

4.2 Conflicts of Interest

The DMC membership has been restricted to individuals free of apparent significant conflicts of interest. The source of these conflicts may be financial, scientific or regulatory in nature. Thus, neither study investigators nor individuals employed by the sponsor, nor individuals who might have regulatory responsibilities for the trial products, are members of the DMC.

The DMC members should not own stock in the companies having products being evaluated by the clinical trial. The DMC members will disclose to fellow members any consulting agreements or financial interests they have with the sponsor of the trial, with the contract research organization (CRO) for the trial (if any), or with other sponsors having products that are being evaluated or having products that are competitive with those being evaluated in the trial. The DMC will be responsible for deciding whether these consulting agreements or financial interests materially impact their objectivity.

The DMC members will be responsible for advising fellow members of any changes in these consulting agreements and financial interests that occur during the course of the trial. Any DMC members who develop significant conflicts of interest during the course of the trial should resign from the DMC.

DMC membership is to be for the duration of the clinical trial. If any members leave the DMC during the course of the trial, the sponsor, in consultation with the SC, will promptly appoint their replacements.

5. TIMING AND PURPOSE OF THE DMC MEETINGS

5.1 Organizational Meeting

The initial meeting of the DMC will be an Organizational Meeting. It will be held during the final stages of protocol development, to provide advisory review of scientific and ethical issues relating to study design and conduct, to discuss the standard operating procedures for the role and functioning of the DSMB, and to discuss the format and content of the Open and Closed Reports that will be used to present trial results at future DSMB meetings

The Organizational Meeting will be attended by the DMC, and by representatives of the sponsors, the lead trial investigators, the independent statistician, and the CRO. The DMC will be provided the drafts of the clinical trial protocol, the Statistical Analysis Plan, the DMC Charter, and the current version of the case report forms. The DMC will also receive the initial draft of the Open and Closed Reports.

5.2 Early Safety/Trial Integrity Reviews

One or more 'Early Safety/Trial Integrity Reviews' will be held during the early stage of protocol enrollment, to review early safety information, to review factors relating to quality of trial conduct, and to ensure proper implementation of procedures to reassess the sample size.

[***The DMC Charter should indicate the expected frequency of these meetings and specification of venue, specifically indicating whether these reviews would be held in person or by teleconference.***]

5.3 Formal Interim Analysis Meetings

One or more 'Formal Interim Analysis' meetings will be held to review data relating to treatment efficacy, patient safety and quality of trial conduct.

[***The DMC Charter should indicate the expected frequency of these meetings and specification of venue, specifically indicating whether these reviews would be held in person or by teleconference. The Charter should also indicate the expected attendees.***]

6. PROCEDURES TO ENSURE CONFIDENTIALITY & PROPER COMMUNICATION

To enhance the integrity and credibility of the trial, procedures will be implemented to ensure the DMC has sole access to evolving information from the clinical trial regarding comparative results of efficacy and safety data, aggregated by treatment arm. An exception will be made to permit access to the independent statistician who will be responsible for serving as a liaison between the database and the DMC. The study's Medical Monitor will be provided immediate access on an ongoing basis to patient-specific information on serious adverse events (AEs) to satisfy the standard requirement for prompt reporting to the regulatory authorities.

At the same time, procedures will be implemented to ensure proper communication is achieved between the DMC and the trial investigators and sponsor. To provide a forum for exchange of information among various parties who share responsibility for the successful conduct of the trial, a format for Open Sessions and Closed Sessions will be implemented. The intent of this format is to enable the DMC to preserve confidentiality of the comparative efficacy results while at the same time providing opportunities for interaction between the DMC and others who have valuable insights into trial-related issues.

6.1 Closed Sessions

Sessions involving only DMC membership and the independent biostatistician who generated the Closed Reports (called Closed Sessions) will be held to allow discussion of confidential data from the clinical trial, including information about the relative efficacy and safety of interventions. In order to ensure that the DMC will be fully informed in its primary mission of safeguarding the interest of participating patients, the DMC will be unblinded in its assessment of safety and efficacy data.

At a final Closed Session, the DMC will develop a consensus on its list of recommendations, including that relating to whether the trial should continue.

6.2 Open Session

In order to allow the DMC to have adequate access to information provided by the sponsor, by study investigators, or by members of the regulatory authorities, a joint session between these individuals and DMC members (called an Open Session) will be held between the Closed Sessions. This session gives the DMC an opportunity to query these individuals about issues that have arisen during their review in the initial Closed Session. With this format, important interactions are facilitated through which problems affecting trial integrity can be identified and resolved. These individuals will either be present at the DMC meeting or be provided a telephone link.

6.3 Open and Closed Reports

For each DMC meeting, Open and Closed Reports will be provided (see Section 8 for outlines of the content of these reports). Open Reports, available to all who attend the DMC meeting, will include data on recruitment and baseline characteristics, and pooled data on eligibility violations, completeness of follow-up and compliance. The statistician (***specify primary trial statistician or independent statistician***) will prepare these Open Reports.

Closed Reports, available only to those attending the Closed Sessions of the DMC meeting, will include analyses of primary and secondary efficacy endpoints, subgroup and adjusted analyses, analyses of AEs and symptom severity, analyses of laboratory data, and Open Report analyses that are displayed by intervention group. These Closed Reports will be prepared by an unblinded independent biostatistician, with assistance from the study biostatisticians in a manner to allow them to remain blinded.

The Open and Closed Reports should provide information that is accurate, with follow-up that is complete to within two months of the date of the DMC meeting. The Reports should be provided to DMC members approximately three days prior to the date of the meeting.

6.4 Minutes of the DMC Meeting

The DMC will prepare minutes of their meetings. Two sets will be prepared: the Open Minutes and the Closed Minutes.

The Open Minutes will describe the proceedings in the Open Session of the DMC meeting, and will summarize all recommendations by the DMC. Since these minutes will be circulated immediately to the sponsor and to lead study investigators, it is necessary that these minutes do not unblind the efficacy and safety data if the DMC is not recommending early termination.

The Closed Minutes will describe the proceedings from all sessions of the DMC meeting, including the listing of recommendations by the Committee. Because it is likely that these minutes will contain unblinded information, it is important that they are not made available to anyone outside the DMC. Rather, copies will be archived by the DMC chair and by the statistician preparing the interim reports, for distribution to the sponsor, lead investigators, and regulatory authorities at the time of study closure.

The sponsors will provide a complete collection of Open and Closed Minutes to regulatory authorities at the time of new drug applications and biologic licensing applications.

6.5 Recommendations to the Steering Committee (SC)

At each meeting of the DMC during the conduct of the trial, the DMC will make a recommendation to the Steering Committee to continue or to terminate the trial. This recommendation will be based primarily on safety and efficacy considerations and will be guided by statistical monitoring guidelines defined in this Charter.

(The Steering Committee will be comprised of the sponsor's study team and lead study investigators, who jointly will have responsibility for the design, conduct and analysis of the clinical trial. The SC will be a multidisciplinary group of approximately seven to ten members who, collectively, have the scientific, medical and clinical trial management experience to conduct and evaluate the trial.)

The SC is jointly responsible with the DMC for safeguarding the interests of participating patients and for the conduct of the trial. Recommendations to amend the protocol or conduct of the study made by the DMC will be considered and accepted or rejected by the SC. The SC will be responsible for deciding whether to continue or to stop the trial based on the DMC recommendations.

The DMC will be notified of all changes to the protocol or to study conduct. The DMC concurrence will be sought on all substantive recommendations or changes to the protocol or study conduct prior to their implementation.

The SC may communicate information in the Open Report to senior management and may inform them of the DMC-recommended alterations to study conduct or early trial termination in instances in which the SC has reached a final decision agreeing with the recommendation. The SC will maintain confidentiality of all information it receives other than that contained in the Open Reports until after the trial is completed or until a decision for early termination has been made.

7. STATISTICAL MONITORING GUIDELINES

[***The DMC Charter should specify the statistical monitoring procedures that will be used by the DMC to guide their recommendations regarding termination or continuation of the trial. These procedures should include guidelines relating to early termination for benefit, as well as guidelines for termination when evidence indicates the experimental intervention has an unfavorable benefit-to-risk profile.***]

[***The DMC may also be asked to ensure procedures are properly implemented to adjust study sample size or duration of follow-up to restore power, if protocol specified event rates are inaccurate. If so, the algorithm for doing this should be clearly specified.***]

8. CONTENT OF THE DMC'S OPEN AND CLOSED REPORTS

8.1 Open Statistical Report: An Outline

- One-page outline of the study design, possibly with a schema
- Statistical commentary explaining issues presented in Open Report figures and tables
- DMC monitoring plan and summary of Open Report data presented at prior DMC meetings.
- Major protocol changes
- Information on patient screening

- Study accrual by month and by institution
- Eligibility violations
- Baseline characteristics (pooled by treatment regimen)
  - Demographics
  - Laboratory values and other measurements
  - Previous treatment usage and other similar information
- Days between randomization and initiation of treatment
- Adherence to medication schedule (pooled by treatment regimen)
- Attendance at scheduled visits (pooled by treatment regimen)
- Reporting delays for key events (pooled by treatment regimen)
- Length of follow-up data available (pooled by treatment regimen)
- Participant treatment and study status (pooled by treatment regimen)

8.2 Closed Statistical Report: An Outline

- Detailed statistical commentary explaining issues raised by Closed Report figures and tables (by coded treatment group, with codes sent to DMC members by a separate mailing)
- DMC monitoring plan and summary of Closed Report data presented at prior DMC meetings
- Repeat of the Open Report information, in greater detail by treatment group
- Analyses of primary and secondary efficacy endpoints
- Subgroup analyses and analyses adjusted for baseline characteristics
- Analyses of adverse events and overall safety data
- Analyses of lab values, including basic summaries and longitudinal analyses
- Discontinuation of medications
- Information on crossover patients

Index

Note: Figures and Tables are indicated by *italic page numbers*; the abbreviation 'DMC' means 'data monitoring committee'

Statistics in Practice

Human and Biological Sciences

Brown and Prescott – Applied Mixed Models in Medicine
Ellenberg, Fleming and DeMets – Data Monitoring Committees in Clinical Trials: A Practical Perspective
Marubini and Valsecchi – Analysing Survival Data from Clinical Trials and Observation Studies
Parmigiani – Modeling in Medical Decision Making: A Bayesian Approach
Senn – Cross-over Trials in Clinical Research
Senn – Statistical Issues in Drug Development
A. Whitehead – Meta-analysis of Controlled Clinical Trials
J. Whitehead – The Design and Analysis of Sequential Clinical Trials, Revised Second Edition

Earth and Environmental Sciences

Buck, Cavanagh and Litton – Bayesian Approach to Interpreting Archaeological Data Webster and Oliver – Geostatistics for Environmental Scientists

Industry, Commerce and Finance

Aitken – Statistics and the Evaluation of Evidence for Forensic Scientists
Lehtonen and Pahkinen – Practical Methods for Design and Analysis of Complex Surveys
Ohser and Mücklich – Statistical Analysis of Microstructures in Materials Science